Reclaiming the Atmospheric Commons

American and Comparative Environmental Policy
Sheldon Kamieniecki and Michael E. Kraft, series editors

For a complete list of books in the series, please see the back of the book.

Reclaiming the Atmospheric Commons

The Regional Greenhouse Gas Initiative and a New Model
of Emissions Trading

Leigh Raymond

The MIT Press
Cambridge, Massachusetts
London, England

This book was set in Stone by the MIT Press. Printed and bound in the United States of America.

Library of Congress Cataloging-in-Publication Data

Names: Raymond, Leigh Stafford, author.
Title: Reclaiming the atmospheric commons : the Regional Greenhouse Gas
    Initiative and a new model of emissions trading / Leigh Raymond.
Description: Cambridge, MA : MIT Press, [2016] | Series: American and
    comparative environmental policy | Includes bibliographical references and index.
Identifiers: LCCN 2016001186| ISBN 9780262034746 (hardcover : alk. paper) |
    ISBN 9780262529303 (pbk. : alk. paper)
Subjects: LCSH: Emissions trading--Northeastern States. | Environmental
    policy--Northeastern States. | Environmental policy--United States--States.
Classification: LCC HC110.P55 R37 2016 | DDC 363.738/740974--dc23 LC record
available at http://lccn.loc.gov/2016001186

10 9 8 7 6 5 4 3 2 1

Contents

Series Foreword

As the political debate over climate change finally shows real signs of maturing and the movement to take action gains momentum, at least at the state and local levels, increasing attention is likely to be paid to the promise of different policy strategies. Public policy scholars and political scientists know well that it is necessary but by no means sufficient to improve scientific understanding of the causes and consequences of climate change. It is also not enough to build public and policy-maker knowledge of these phenomena, and reach some consensus on the need for governmental intervention. Once the science is deemed to be persuasive enough to act, policy makers also need to know which possible actions to reduce greenhouse gas (GHG) emissions are likely to work best over time, be reasonable in terms of costs imposed on the economy, and attract enough public support to be politically feasible. In short, we need to draw from the best policy analysis and studies of the policy process to guide these decisions.

In the United States, the battle over climate change policy in recent years has been fought with often intense political polarization and partisan gridlock, especially at the national level. It is particularly important, then, to explore policy options, such as market-based approaches, that potentially could attract support across the political spectrum and thus help build a lasting foundation for climate change policy. In 2010, a national cap-and-trade program appeared to meet those conditions, but ultimately the US Congress failed to approve it. Since that time, partisan differences on climate change have substantially widened and made political agreement all but hopeless.

Yet one of the most remarkable features of the US political system is its robust division of power between the federal government and states. Individual states, acting alone or in concert with their neighbors, may adopt innovative policies that are not feasible at the national level. In the case of climate change, careful examination of state-level (and perhaps even municipal) policy experimentation may well yield important insights into which policies hold the most promise, and what kinds of political processes can facilitate their adoption and successful implementation. It is likely that stakeholder engagement, among other approaches, will need to play a role in this regard.

Leigh Raymond's *Reclaiming the Atmospheric Commons* takes on that challenge by analyzing the Regional Greenhouse Gas Initiative (RGGI) cap-and-trade program agreed to by a group of northeastern states. He offers the most thorough study of RGGI to date. In particular, Raymond explores the adoption of the idea of auctioning emissions allowances and making power generators pay for emitting GHGs into the atmosphere, which he characterizes as a revolution in the realm of environmental policy. RGGI's approval constitutes a major case of rapid policy change, and Raymond examines how it came about, especially the role that policy entrepreneurs played in its adoption. He focuses on the processes of issue framing and agenda setting to analyze the idea of creating a new "public benefit" within cap-and-trade policy of broadly distributed benefits that flow from the auction revenues.

It is this reframing of the issue of allocation as a "public entitlement" that lies at the center of Raymond's analysis. He argues that environmentalists successfully promoted the frame as a way to underscore the important and tangible benefits to citizens from auctioning emissions allowances. These public benefits included subsidies for energy efficiency improvements and even direct rebates to utility ratepayers. The Citizens' Climate Lobby has made a similar argument in its recent efforts to promote a bipartisan national climate change policy grounded in the adoption of a revenue-neutral carbon tax whose proceeds would be fully rebated to taxpayers. The group believes that the rebated revenues will make the idea of a carbon tax broadly attractive even at a time of sharply divided opinion on climate change policy.

Aside from the fascinating case of RGGI, Raymond contends that the public benefit frame also has influenced the advancement of emissions trading in the European Union and California's cap-and-trade program. Hence, it is crucial to understand what role it played in RGGI, and what it might do to further climate change policy in many other venues at a time when the world's nations appear to be ready to undertake serious policy steps. As the first in-depth treatment of this development in climate change policy, the book will receive thoughtful attention among both scholars and policy makers. It should stimulate new research on the little-understood processes of policy change, and especially the role of "normative reframing" as one element within those processes.

The book illustrates well our purpose in the MIT Press series in American and Comparative Environmental Policy. We encourage work that examines a broad range of environmental policy issues. We are particularly interested in volumes that incorporate interdisciplinary research, and focus on the linkages between public policy and environmental problems and issues both within the United States and in cross-national settings. We welcome contributions that analyze the policy dimensions of relationships between humans and the environment from either a theoretical or empirical perspective.

At a time when environmental policies are increasingly seen as controversial, and new and alternative approaches are being implemented widely, we especially encourage studies that assess policy successes and failures, evaluate new institutional arrangements and policy tools, and clarify new directions for environmental politics and policy. The books in this series are written for a wide audience that includes academics, policy makers, environmental scientists and professionals, business and labor leaders, environmental activists, and students concerned with environmental issues. We hope they contribute to the public understanding of environmental problems, issues, and policies of concern today, and also suggest promising actions for the future.

Sheldon Kamieniecki, University of California at Santa Cruz
and
Michael Kraft, University of Wisconsin at Green Bay
Coeditors, American and Comparative Environmental Policy Series

Acknowledgments

This book represents the collision of two interests that I developed about ten years ago: the prospects for auctioning emissions allowances in cap-and-trade programs, and the ability of policy theories to explain sudden policy change. As early as the publication of my 2003 book on market-based environmental policies, I was intrigued by the seemingly unthinkable idea that a cap-and-trade policy might one day make emitters *pay* for their valuable emissions allowances. Despite understandable skepticism of the idea by many colleagues, I kept an eye out for potential examples. Sure enough, a few years later I noted with great interest the decision of the Regional Greenhouse Gas Initiative (RGGI) to auction nearly all its emissions allowances.

During this time, I had also been contemplating the struggles of leading policy theories to explain sudden, nonincremental policy change. In considering those theoretical issues, I saw the connection to the surprising policy change that had just occurred in RGGI. The result is this book, which leverages the RGGI case and subsequent events in climate policy to also improve our understanding of how norms can be a powerful political tool in creating dramatic policy change, even in the face of opposition by powerful interests.

It is fair to say that without the support of two senior scholars, this book is unlikely to have happened. Denny Ellerman is one of the "deans" of emissions trading—someone who has literally written the book (and several more) on the subject. His early interest in my work and periodic feedback on my ideas has been incredibly valuable. Similarly, Barry Rabe is another leader in the field of climate policy research, especially state and subnational climate policies, and his interest in and comments on my work over the years have been invaluable. While I would not suggest that Denny or Barry would agree with everything I've concluded in this volume, I do want to thank them for making this work possible in many ways.

Barry in particular helped spark this research agenda with the invitation to present at the University of Virginia Miller Center's 2008 conference on Greenhouse Governance, where I delivered a talk and subsequent book chapter that drew attention to what I was calling the "RGGI revolution" even then. Those who know Barry will realize this is par for the course—few scholars are as generous with their time and energy as he is in working with and supporting his colleagues, and I'm fortunate to have gotten to interact with him regularly about the never-ending story of state and federal climate policies. Barry also read and commented on an early draft of the book's chapter on RGGI. Other scholars interested in RGGI who have provided excellent feedback include Bruce Huber (who also provided detailed comments on a preliminary draft of the chapter on RGGI) at the University of Notre Dame, and Roger Karapin at the City University of New York.

Like all of my research, this manuscript benefited tremendously from the detailed and careful comments of many of my colleagues at Purdue University. The Purdue political science faculty is a remarkably collegial and interdisciplinary group, and I'm lucky to have many of its members as regular readers of drafts of my work. In particular, Patricia Boling, Rosie Clawson, Aaron Hoffman, and Laurel Weldon meticulously read drafts of one or more chapters of the manuscript. Their feedback and encouragement improved the final product greatly, and I am grateful for their time and energy as well as their outstanding company. Laurel Weldon in particular has been a strong supporter of the book, and I can only hope that my work on norms and policy change rises to the level of her own research in this area.

Finally, I thank the three reviewers at the MIT Press as well as the Environmental Studies acquisitions editor, Beth Clevenger, and American and Comparative Environmental Policy Series editors, Sheldon Kamieniecki and Michael Kraft, for their interest in and careful reading of the manuscript. I'm also grateful to Clay Morgan at the MIT Press for his early expression of interest in the book idea. Despite all this excellent feedback and commentary, of course any mistakes that remain in the manuscript are my own.

I also need to acknowledge Ximena Arriaga, Ann Marie Clark, Dan Kelly, and Laurel Weldon, who helped organize an interdisciplinary, international workshop in 2012 at Purdue on norms and intractable global problems. This workshop brought together experts on norms and informal institutions from a variety of disciplines working in a wide array of policy contexts, and expanded my understanding of the influence of norms in social and political life tremendously. I thank my co-organizers for their efforts and insights as well as the other invited guests and audience members at the event.

Like many books, drafts of different chapters of this one were also presented at several conferences or workshops, including the University of Michigan in 2010, the Midwest Political Science Association research conferences in 2012 and 2014, and the Purdue workshop in Public Policy and Political Theory in 2014. I am grateful to the audience members in all of those sessions for their attention and their comments and questions.

Although solitary in many ways, completing a book of this nature also required the help of many other individuals. I offer particular thanks to all of the very busy people who took time to speak with me at length about their experiences with RGGI and other emissions trading programs, sometimes completing a second or third follow-up interview when I had additional questions. I especially want to thank Franz Litz, former cochair of the RGGI Staff Working Group (SWG), for providing me with access to many of his personal notes and records of the RGGI process: an invaluable source of information. In addition, William Shobe supplied a similar level of access to his own records of the little-known story of the policy decision to auction nitrogen oxide (NO_x) allowances in Virginia in 2004, for which I am also grateful. I also want to thank both the Purdue College of Liberal Arts and Purdue Climate Change Research Center, which provided important financial support for this research.

Mo Lifton is a great resource at Purdue for graphic design, and was responsible for creating print-ready versions of several figures for chapter 2. I also thank Paul Hibbard and Lauren Zawilenski for permission to reprint their graphics or photos in the text. Finally, I appreciate the patience and support Miranda Martin and Virginia Crossman at the MIT Press provided during the manuscript preparation and submission process.

In addition, a number of students offered crucial research support on the project over the past six years, ranging from then high school senior Angela Filley, to undergraduate research assistants Brandon Dittman and Allison Roberts, to graduate research assistants Heather Cann, Adam Dahl, and Andrew Tuholski. It is a pleasure to include students in the research process, and their various contributions are evident throughout the text.

More important, I thank my wife, Teresa, who provides the support that keeps me going through the ups and downs of pursuits such as this one. For her role in making this book possible, and so much else, I will always be thankful.

1 Introduction

In 2008, a group of northeastern states achieved what had been thought a political impossibility: a policy that forced polluters to pay the public for their emissions. Inaugurating their new emissions trading program, the Regional Greenhouse Gas Initiative (RGGI), these ten states sold approximately 12.5 million emissions "allowances," or legal rights to emit a ton of pollution, at a clearing price of $3.07 per ton, raising approximately $38.5 million for investments in energy efficiency and other public benefits (RGGI 2015c). Through 2014, RGGI (ibid.) states had conducted a total of 29 auctions, selling more than 785 million allowances and raising more than $2.2 *billion* in revenue, most of which was returned to the public. More broadly, the decision to auction allowances for greenhouse gases (GHGs) in the entire United States could result in hundreds of billions of dollars of revenue, or more than a hundred times the value of previous national emissions trading programs for nitrogen oxides (NO_x) and sulfur dioxide (SO_2) (Palmer 2010).

This decision to force companies to pay for the right to use the "atmospheric commons" and dedicate those revenues to tangible public benefits was a revolutionary change in the decades-long history of market-based environmental policies. At the start of the RGGI process in 2003, anyone predicting such an outcome would have been accused of living in an environmentalist fantasyland (e.g., Hahn 1989; Keohane, Revesz, and Stavins 1998; Ellerman et al. 2000, chap. 2). Although economists had promoted the idea of auctioning emissions allowances under "cap-and-trade" policies such as RGGI's, their suggestions had long fallen on deaf ears. Instead, governments gave these valuable assets away to companies for free, generally in rough proportion to their recent emissions. This practice of "grandfathering" allowances was consistent with theories of interest group politics describing the durability of policies creating concentrated benefits for a few powerful interests and widely distributed costs for the public (e.g., Wilson 1989).

It was also consistent with prominent criticisms of emissions trading policies for providing private profits rather than public benefits (Klein 2014; Bachram 2004). According to this perspective, cap-and-trade policies allow private firms to increase their profits by buying and selling valuable "rights to pollute" at the expense of the public and environment. The decision to auction all emissions allowances, and dedicate that revenue to programs benefiting the public at large, was in stark contrast to this image of emissions trading. Rather than continuing to give away the right to use the atmosphere, auctions articulated an alternative vision that this resource was owned by the public. In this sense, RGGI's policy designers "reclaimed the atmospheric commons" for citizens, contrary to the standard practice of providing the valuable use of that resource for free to private firms.

Given the United States' withdrawal from international climate negotiations related to the Kyoto Protocol in 2001, it was hard to imagine that a group of US states would be successful at enacting any new policy to limit GHG emissions at the time the RGGI process began. Trying to *also* make generators pay for their emissions for the first time as a part of that policy seemed like a classic case of overreach. As Franz Litz (2011), a high-ranking administrator in the New York Department of Environmental Conservation, later put it, he and his peers in other states did not initially believe there was "even a remote possibility" that auctions could be a part of RGGI.

Despite this skepticism, a small group of environmentalists decided to promote the idea. Five years later, RGGI states had required electricity generators to pay for more than 90 percent of their GHG emissions—a result that surprised even those who were deeply involved in the RGGI design process. Europeans who had developed their own emissions trading program around the same time period, for example, were reportedly "astonished" that RGGI was able to enact nearly 100 percent auctioning (Burtraw 2011). In short, at a time when other emissions trading programs continued to give away these valuable rights, RGGI was able to auction nearly all of them.

Some observers have already identified this sudden adoption of auctions as the most important innovation in RGGI (Tietenberg 2010; Cook 2010; Rabe 2010), or in the words of one participant (Murrow 2010), as "the most significant thing that came out of the whole RGGI process." Given the amount of money at stake, and the fact that RGGI has served as a model for other cap-and-trade policies since 2008, this outcome offers crucial implications for future climate policies. Given the powerful interests opposed to this new "cap-and-auction" approach and lack of any clear political precedent, it also offers important new ideas for our understanding of sudden policy change in general. Yet it is a change that few have been fully able to explain to date.

This book will show how advocates accomplished this feat through a new strategic reframing of the issue—one that stressed the public ownership of the atmospheric commons. By reframing the issue in terms of alternative norms requiring polluters to pay for their use of such a resource and emphasizing the need for a widely distributed "public benefit" from any such private use, environmental advocates were able to gain sufficient political support for auctioning emissions allowances where previous efforts had fallen short.

Thus, RGGI's adoption of auctions was a "revolutionary" change in climate policy—one that asserted stronger public ownership rights over a shared atmospheric resource and rejected long-standing assumptions about free allocations to existing resource users. Despite growing political hostility to policies addressing climate change since RGGI's initiation in 2008, other programs have relied on variations of the same public benefit arguments to justify new or expanded cap-and-trade designs in a wide range of locations, including some of the largest cap-and-trade policies in the world in California and the European Union. In these subsequent policy debates, policy designs that have stuck closer to the RGGI model have tended to fare better politically, while those deviating from RGGI's example of limiting the use of auction revenue for broadly distributed public benefits have fared worse. Although RGGI's initial emissions reductions were modest, they introduced an entirely new model for conceptualizing climate change policy—a model that has since been used successfully in a range of additional settings.

This is not to argue that the new public benefit framing was the only factor leading to the adoption of auctions in RGGI, or that such a reframing will work in every political context. Rather, it is to assert that the new public benefit framing played a critical role in the surprising and significant decision to adopt auctions in RGGI, and appears to be one of the more promising strategies for promoting new climate policies in a variety of settings even as partisan opposition to action on climate change increases. In addition, the success of public benefit framing in RGGI and other cases offers important insights toward a new theoretical model of sudden policy change in general.

"Normative Reframing" and Sudden Policy Change

Through their reframing strategy, RGGI's auction advocates took advantage of the distinctive power of norms to spur dramatic policy change. Generally defined as standards of behavior appropriate for a given identity (Finnemore and Sikkink 1998), *norms* are the informal rules that structure a major part of human thinking and behavior (Ostrom 1998; Sripada and Stich 2007). Because they are able to inspire behavior or attitudes in conflict with powerful motivations of self-interest (Henrich et al. 2004; Ostrom 1998; Axelrod 1986), norms are especially important levers for social change (Raymond et al. 2014). Recognizing this fact, advocates have used norm-based strategies to promote political and social change on issues ranging from human rights (Sikkink 2011) to domestic violence (Htun and Weldon 2012) to health care (Lynch and Gollust 2010; Kingdon 2003), among many others.

One crucial but previously unrecognized strategy for using norms to promote policy change is what this book refers to as *normative reframing*, or the strategic use of issue frames to portray an issue in terms of an alternative norm. Grounded in the fields of political communication and political psychology, *issue frames* are generally defined as conceptualizations of a problem that emphasize different aspects of the issue, thereby suggesting different policy responses (Entman 1993). It is well established that issue frames can substantially influence public attitudes toward policy choices, and that frames vary in their "strength" or ability to change those attitudes (Chong and Druckman 2007). Given the importance of norms in shaping many public attitudes and behaviors, frames portraying an issue in terms of strong social norms should offer unique leverage in shifting public and elite policy attitudes. As such, this book argues that one can better explain sudden policy change through greater attention to the application of normative frames in particular to an issue.

A normative frame's ability to shape policy attitudes is conceptualized here as a product of its normative "force" and "fit." By normative *force*, the theory refers to how much influence a norm has over attitudes and behaviors in a given society. Some norms are central to a particular society and carry severe social penalties for violation, while others are violated more easily. All things being equal, frames portraying an issue in terms of a norm with greater normative force will be more effective in generating support for a policy option.

At the same time, norms are often ambiguous in terms of their applicability (Richerson and Henrich 2009). Given this ambiguity, individuals will perceive some norms as applying to or "fitting" an issue more clearly than others, especially when forced to reflect on the potential relevance of two competing norms. It is possible in this sense to define normative *fit* as the perceived applicability of a norm to a specific issue. Efforts to reframe a basic level of income as "new property" (e.g., Reich 1964) subject to strong norms requiring respect for private ownership, for example, may have failed politically due to a relatively poor fit between norms of ownership (typically applied to land or other "things") and the more intangible concept of a minimum income. In this respect, a normative frame with a weaker normative fit uses a norm most people decide is not clearly applicable to a particular issue. Frames portraying an issue in terms of a norm with a better normative fit, by contrast, will be more effective in generating support for a policy option.

Normative reframing is a three-step process. First, change advocates must identify the primary norm or norms supporting the policy in place. Second, they must highlight this existing normative frame and persuade people of its weak fit with the issue in order to destabilize the status quo. Finally, change advocates then reframe the issue in terms of alternative norms they believe will be perceived to fit the issue better. Although the evaluation of a norm's fit with an issue is a matter of personal judgment, people in a given society frequently agree on which norms are more or less appropriately applied to many situations, such as the idea that the norm of tipping does not fit well with the situation of having dinner at a friend's home (Ariely 2009). In this sense, the range of norms that can be persuasively applied to a given issue is limited.

Reframing an issue in terms of an alternative norm should in turn make a new policy more politically viable due to its improved consistency with the alternative norm being applied. In addition, by studying the perceived fit between normative frames and the policies they support, policy theorists and advocates should be able to better identify and predict which policies are more likely to experience sudden policy change in the future. Policies supported by norms that are perceived as having a poor fit with the issue are more vulnerable to sudden policy change than those supported by norms seen as having a stronger fit.

Normative Reframing and the RGGI Revolution

Normative reframing is vital to explaining the sudden policy change in the RGGI case. Environmental policy entrepreneurs noted the poor fit between a "Lockean" norm justifying private rights to a natural resource based on beneficial labor and the policy of allocating free emissions allowances. In this sense, auction proponents recognized that grandfathering allowances to current polluters effectively rewarded the *harmful* behavior of polluting: a strained application of a norm rewarding *beneficial* prior use. By drawing attention to the weak normative foundation for grandfathering emissions allowances, or "foregrounding" the weak fit of the Lockean norm in this case, advocates undermined political support for the practice. As an alternative, auction supporters reframed allowance allocation as a process of granting private companies limited rights to use a public resource (the atmosphere), in accordance with alternative norms suggesting that such policies must benefit all citizens in a broad manner. This new *public benefit framing*, combined with other factors, made the previously unthinkable idea of selling emissions allowances politically viable.

Two norms were equally prominent in the new public benefit frame promoted by RGGI's auction advocates. First, environmentalists described auctions as justified by the long-standing norm that "polluters should pay" for their emissions. Previously unsuccessful efforts to promote auctions indicated, however, that the "polluter pays" idea was insufficient by itself. Because emitters threatened to pass the higher cost of auctioned allowances on to consumers, for example, simply following this norm threatened to raise energy prices for the public, making it less effective.

For this reason, it was crucial to emphasize a second egalitarian norm requiring a broad public benefit from the private use of this public resource. Making polluters pay was fine, but environmentalists eventually recognized it was as or more important to provide tangible benefits from those payments to the public at large. Their successful new auction frame for RGGI accordingly stressed the need for tangible and widely distributed public benefits from any private use of this public resource. Based on this new public benefit frame, advocates promoted auction policy designs that used revenue to help consumers reduce their energy bills. This in turn allowed auctions to gain significant new political traction, ultimately surpassing the expectations of even the idea's strongest advocates. Taken together, the public benefit frame and associated policy design can be thought of as a new "public benefit model" for climate policy that was remarkably successful in RGGI, and has shown promise in other political contexts.

Implications of Normative Reframing and the Public Benefit Model

The success of the public benefit model in RGGI has notable implications for future environmental policies. In the face of growing political opposition to all climate change policies, cap and trade *with auction* is now one of the most common policies for addressing climate change around the globe. Post-RGGI developments also suggest that the key climate policy conflict has become distributional—who pays, and who benefits, from the policy. Indeed, the expansion of the public benefit frame suggests that the future of climate policy increasingly will hinge on how policy designers define and allocate the public benefits of the private use of the atmospheric commons. These implications are vital to understanding future climate policy developments in the United States and globally, including the development of state emissions policies under the US Environmental Protection Agency's (EPA) Clean Power Plan, and China's forthcoming national cap-and-trade program for GHG emissions.

One example of this conflict is currently playing out over auction revenue in California, which implemented its own cap-and-trade-with-auction policy under its 2006 Global Warming Solutions Act (AB 32). Unlike their RGGI counterparts, California lawmakers did not specify the intended uses for auction revenue beyond a general need to support the goals of AB 32. As a result, political conflict over competing proposals for spending auction revenue has threatened the political fortunes of the program. A similar story has played out in other jurisdictions, including major conflicts over the use of auction revenue in the EU Emissions Trading System (ETS) as well as failed proposals for national cap-and-trade policies in the United States and Australia. In this regard, the decision about how to allocate benefits from future climate policies, and especially from any allowance auction revenue, may become a more pivotal question for policy designers than long-running debates over the science of climate change.

Emerging conflicts over the disposition of auction revenue evoke a more long-standing conflict in US and European culture over limits on the government's discretion to allocate public resources. In asserting public ownership of the atmosphere, for instance, RGGI's designers emphasized the need to return the value of that resource use *to all citizens* in the form of widely available subsidies for energy efficiency improvements or energy bill rebates. This is consistent with the English common law doctrine of the public trust, which stresses equal public access to and benefit from public resources. Traditionally applied to waterways and submerged lands, but more recently expanded to other natural resources, the public trust doctrine offers a conceptualization of public ownership that means something closer to "owned directly by the people." An alternative view defines public ownership more broadly as "owned by the government," thereby giving elected officials greater freedom to spend auction revenue on a wider variety of goals. Some cap-and-auction proposals after RGGI have adhered more to this "government ownership" approach to allocating auction revenue, often generating greater political controversy as a result. How the tension between these different understandings of public ownership is resolved will be a vital question for future climate change policies, especially those using cap and trade.

Despite this important question, events since RGGI's enactment suggest that the public benefit model is likely to have staying power, and that it offers a promising path forward for policies to address climate change and other environmental challenges related to shared common resources. Although norms vary from society to society, and other factors will shape a policy's political fortunes besides its normative framing, policies that hew more closely to the public benefit model first used in RGGI seem to have fared better politically in a variety of contexts. More generally, the evidence indicates that the public benefit model has changed the terms of the debate over regulating air pollution in durable ways.

The theory of normative reframing used to explain RGGI also offers important implications for theories of policy stability and change. Political scientists have struggled to generate theories that predict or explain sudden policy changes. Building on existing policy theories such as punctuated equilibrium (PE) (Baumgartner and Jones 1993; Jones and Baumgartner 2005, 2012), normative reframing provides a new pathway for explaining major policy change, even in the face of opposition from powerful vested interests. This is a significant step beyond policy theories emphasizing the power of vested interests to maintain the policy status quo by controlling political "subgovernments" (e.g., Lowi 1979; Wilson 1989). It also helps move explanations of sudden policy change beyond "shocks" to policy systems that are themselves unpredictable (e.g., Weible, Sabatier, and McQueen 2009).

PE theorists describe how change advocates promulgate new "policy images," thereby shifting the issue's political venue as well as the composition of political coalitions (Baumgartner, Jones, and Mortensen 2014). These policy images resemble issue frames in that they represent alternative conceptualizations of an issue. Normative reframing builds on this idea of policy images while drawing on the distinctive power of norms in shaping attitudes and behavior. By promoting an alternative normative frame for an issue, change advocates propose a new policy image with unique influence—one that applies a new norm to the issue, thereby suggesting new policy arrangements. In this manner, change advocates leverage the distinctive power of social norms in both challenging political support for an existing policy and generating political support for an alternative. This strategy is likely to be effective across many policy domains where prominent norms play a central role, including many environmental policy conflicts.

Normative reframing also offers a way to better identify specific policies that are more or less likely to experience sudden policy change. As noted above, policies supported by norms that many individuals see as having a poor fit with the issue or by norms that are themselves controversial in a society are more vulnerable to policy change. In this respect, the normative reframing theory advanced in this book builds on policy theories such as PE by offering a way to make testable predictions of the likelihood of specific policy changes rather than predictions of the relative prevalence of large versus small policy changes in a given domain over time. Similar to PE theory, it preserves a greater role for agency in our theories of policy change by recognizing how change advocates can identify and destabilize policies using new policy images or issue frames.

Plan of the Book

The book presents its arguments about the power of normative reframing to explain sudden policy change and the political advantages of the public benefit model in the following steps. Chapter 2 reviews existing theories of the policy process, highlighting how such theories continue to struggle to explain sudden policy change. This discussion reviews long-standing theories of how powerful private groups create and maintain policies that serve their interests, dating back to the work of Mancur Olson (1965) and Theodore Lowi (1979). It then describes the unique power of norms in shaping human attitudes and behavior. After reviewing the underutilized role of norms in many leading policy theories, including the advocacy coalition framework, PE theory, and the multiple streams framework, the chapter presents its new theory of normative reframing as a way to improve our understanding of intentional and sudden policy change.

Having presented and contextualized this new theoretical model of policy change, the book explores important precedents to the surprising adoption of allowance auctions in RGGI. Chapter 3 reviews the long history of support by economists for making polluters pay for their emissions, either through an "emissions tax" as promoted by Arthur Pigou (1920), or via a tradable private "property right" to pollute, consistent with the recommendations of Ronald Coase (1960). The chapter also considers early emissions trading policies in the 1970s and 1980s, in which politicians and regulators ignored the recommendations of economists and repeatedly gave away these valuable emissions rights. Next, the chapter describes emergent "seeds of change" for emissions trading policy in the 1990s, including enactment of the landmark 1990 US cap-and-trade program for SO_2, deregulation of electricity markets in the northeastern United States, creation of auctions or severance taxes on other public resources, and promulgation of new public benefit charges on electricity bills in many states that helped consumers lower their energy consumption and electricity bills.

Despite these changes, however, chapter 3 examines how efforts to promote auctions failed in two prominent cases just prior to RGGI. First, it reviews unsuccessful efforts to promote auctions of allowances to existing sources in new emissions trading programs for NO_x in the eastern United States from 1994 to 2004. In addition, the chapter analyzes failed efforts from 1997 to 2003 to authorize auctions of more than a small percentage of allowances in the initial stages of the EU ETS. Both examples serve as important failures for comparison with the adoption of auctions in RGGI, particularly the NO_x program, which involved many of the same actors and causal factors present in the RGGI case.

Chapter 4 reviews the lengthy process by which advocates were able to promote auctions successfully for the first time in the creation of RGGI. The chapter relies on records from every stage of the five-year policy design process, including thousands of pages of public comments, meeting minutes and summaries, stakeholder presentations, and draft policy rules and memoranda. The chapter also relies on the personal notes and records of former officials involved with the program from start to finish as well as detailed interviews with more than thirty key actors in the RGGI design process, including environmentalists, industry and power generator representatives, technical consultants, state agency leaders, and elected officials.

This analysis provides an account of the adoption of auctions in RGGI that differs from previous studies in several key ways. For one, it identifies environmental advocates as those who first conceived and promoted the idea of auctioning allowances, rather than state agency officials or politicians. Second, it describes the significance of the new public benefit frame in building political support for the auction idea as opposed to general frames about promoting economic development. In particular, the chapter describes how the new public benefit frame required using auction revenue to fund programs benefiting the public directly instead of for general government expenses or to favor certain industries. In this respect, it documents the importance of earlier public benefit programs that lowered electricity and heating bills for all ratepayers as a crucial precedent for auctions in RGGI. Finally, the chapter illustrates how auctions were a prominent part of the policy discussion from the outset rather than arising late in the RGGI process.

The chapter pays particular attention to the vital role of the new public benefit frame and its associated policy of dedicating auction revenue to tangible benefits for consumers in making RGGI politically viable. Initially considered an outrageous suggestion, auctions became a linchpin of the strategy for enacting RGGI by helping to protect consumers from expected electricity price increases. This experience is an instructive contrast with failed efforts to promote allowance auctions, such as in the EU ETS or NO_x trading programs, which invoked the polluter pays norm, but failed to incorporate the additional egalitarian norm that completed RGGI's public benefit model. Based on this process-tracing exercise and the comparison with the failed cases described in chapter 3, the new normative frame stressing the importance of an egalitarian distribution of any auction revenue emerges as a critical factor for making the switch to auctioning allowances possible in RGGI.

Chapter 5 then reviews evidence that the public benefit model has continued to influence the climate policy discourse beyond RGGI and may offer the best option for promoting new climate policies in the future. Emissions trading policies to address climate change have faced a more difficult path since RGGI's implementation due to growing conservative opposition to any climate change policies and a global economic recession. The chapter reviews the mixed results of other existing or proposed cap-and-trade programs in the United States and abroad. The pattern is consistent: the further a policy strays from the public benefit model, the more political problems it has had. Although these data remain preliminary, the evidence supports the book's argument that the new public benefit model introduced in RGGI is on more stable political ground than previous policies relying on grandfathering emissions, and therefore is more likely to endure politically. In addition, the chapter portrays ongoing efforts to expand the scope of the public benefit model to include a wider range of benefits, including public health gains from reductions in copollutants associated with GHG emissions. The chapter concludes that future climate policy controversies are likely to turn on distributional policy provisions and the perceived acceptability of different variations of the public benefit frame.

Finally, the chapter describes how normative reframing could help explain a wider range of policy changes. Social scientists already use standard research methods such as content analysis, surveys, focus groups, and in-depth interviewing to identify the frames supporting a particular policy status quo. Building on these efforts, policy scholars should also be able to use similar methods to measure public or elite perceptions of the apparent strength of the norms featured in those frames as well as their perceived fit with the policy at hand. The chapter then shows how scholars and advocates can use these methods to identify public policies that are more vulnerable to sudden policy change in the near future, based on the relative strength and fit of the norms currently supporting them. In this manner, normative reframing offers a mechanism for making more specific predictions of policies prone to sudden change—an important step forward in our theories of policy stability and change in general.

Conclusion

Scientists agree that without quick action, climate change impacts will pose a profound risk to human society. Furthermore, political observers frequently point to the challenge of making major policy changes to address this issue given the substantial economic interests at stake. More generally, observers and theorists of public policy have long noted the difficulty of intentionally creating or even predicting sudden policy change. For all these reasons, it is vital to consider cases like RGGI, where a small group was able to create a major policy innovation on a seemingly intractable problem like climate change. The RGGI case also is critical as an early example of the apparent potential for a new approach to climate policy based on a new vision of the "atmospheric commons" to increase pressure for policy action on this issue. By better understanding the power of the public benefit model as a tool for sudden policy change in RGGI and subsequent cases, we can identify promising strategies for future policy changes to address climate change and many other problems facing the world today.

2 Theorizing Norm-Driven Policy Change

Political scientists have long struggled to explain sudden policy change. Much of this is due to the influence of theories of "interest group pluralism," which predict policy stability when vested interests are served by status quo rules, such as those giving away emissions allowances for free to polluters. More recent policy theories have moved beyond this interest group perspective, but continue to struggle to explain sudden policy changes, often forced to attribute them to other unpredictable shocks such as natural disasters or dramatic elections. Although some of these newer theories, notably PE theory, describe strategies that advocates can use to facilitate major policy shifts, even PE theorists remain pessimistic about the prospects of better predicting any specific policy changes (Baumgartner, Jones, and Mortensen 2014).

The premise of this book is that political scientists can do better on this difficult problem, and that a new focus on norms offers a promising path forward. Although many policy theories recognize the significance of norms, their potential to improve such theories remains underutilized. In particular, policy scholars acknowledge that norms affect policy choice in a variety of contexts and are an important cause of policy stability, but have not fully articulated a theory of how norms can serve as a catalyst for sudden policy *change*. After reviewing existing efforts to theorize nonincremental policy change, the chapter offers an initial account of a theory of norm-driven policy change that could better predict policy change as well as stability in the future. The crux of the theory is based on the idea of *normative reframing*, or using a new issue frame to portray an issue in terms of an alternative norm. The chapter concludes with a conceptual overview of how the process of normative reframing helps explain the sudden change toward auctions in emissions trading policies related to climate change.

Interest-Based Theories of Policy Stability

Long-standing theories describing political choice as dominated by powerful interest groups remain highly influential in our understanding of the
policy process (e.g., Lowi 1979; Wilson 1989; Arnold 1990). Such interest group perspectives are pessimistic about the chances for major policy
change. Distributive policies, such as those allocating emissions rights, are
often cited as paradigmatic examples of interest group politics in which
"iron triangles" and subgovernments prevent substantial change to status quo distributions of resources (McConnell 1966; Bernstein 1955). This
section briefly reviews the traditional interest group perspective on policy
stability and change—the dominant theory used to explain the process of
allocating emissions allowances in cap-and-trade programs to date.

Early policy theories argued that most policy changes are modest and
changing the status quo in politics is extremely difficult. This "incrementalist" perspective offers many reasons for the tendency toward small policy
changes, including cognitive limitations described broadly as "bounded
rationality" (Lindblom 1959). According to this perspective, limits in the
attention and information-processing capacity of policy makers makes
them unable to consider the wider range of policy options recommended
by more idealized models of decision making that stress a "rational-com-
prehensive" approach.

Starting with Olson (1965), an additional understanding of incremental
policy change underscored the costs of collective political action. According to this theory, many policies beneficial to the public will fail due to
insufficient motivation for any individual to invest the effort required to
persuade government to act. The concentrated costs of political organizing and lobbying compared to the widely dispersed public benefits from
such reforms mean they will often lack a champion. Instead, large interest
groups or private entities with a greater stake in the issue will dominate the
policy process due to their higher potential benefits relative to the cost of
organizing politically.

In this manner, the model of interest group pluralism that was once
promoted as a democratic ideal (Truman 1951) was later condemned by
those who argued that only powerful interests were likely to prevail politically, and that less organized or powerful groups in society would not have
their interests adequately represented. Lowi first articulated this critique in

1969, joined by more economically grounded critiques of policy making by theorists such as George Stigler (1972), who saw government responding only to the demand for a "supply" of regulation that met the needs of the most powerful buyers in the political market: well-established business concerns. As summarized by James Q. Wilson (1989), this vision of interest group politics suggests dominance by those with large benefits or costs at stake, and limited influence by those facing only diffuse costs or benefits in the form of slightly higher taxes or small individual gains. On this account, major policy change is rare because it requires overcoming the opposition of powerful groups with substantial vested interests in the policy status quo. Despite its age, the interest group pluralism model of policy making remains central to our understanding of policy stability and change, including most explanations of the process of enacting cap-and-trade programs, as will be reviewed in chapter 3.

Beyond Interests: The Importance of Norms

Alternative theories of the policy process pay greater attention to factors other than interests, or the relative concentration of costs, benefits, and political power, in explaining policy stability and change. Instead, "ideational" factors such as shared norms and beliefs play a more important role, making these theories a valuable expansion of the interest group perspective (Schmidt 2008; Béland and Cox 2011). Before reviewing these alternative theories of policy stability and change, however, it is essential to clarify some key terms and also to describe the distinctive quality of norms as especially significant ideational factors in shaping individual attitudes, collective and individual decisions, and policy choices.

Norms can be conceptualized as one type of belief held by an individual, where *beliefs* are defined as associations of "various characteristics, qualities, and attributes" with a "belief object" (Fishbein and Azjen 2010, 96). *Values*, by contrast, are often defined as evaluative beliefs that are more fundamental and fewer in number, and relate to how preferable a particular "mode of conduct or end state of existence" is for the individual (Feldman 2003, citing Rokeach 1973). Many values have an unconditional or absolute quality that makes them similar to the idea of a shared "moral" standard of appropriate behavior, which some refer to as a "moral norm" (Elster 2007, 104). In this respect, one can think of a moral norm as similar to a shared personal value related to an ethically appropriate standard of behavior. Finally, most political psychologists think of an *attitude* as a narrower type of belief that represents an evaluation of a specific "belief object," such as a public policy, issue, or individual (Feldman 2003). Efforts to strengthen support for various policies are frequently described in terms of the relative influence of different norms, values, and other beliefs (sometimes referred to collectively as "considerations") in shaping an individual's attitudes on the issue (e.g., Nelson, Oxley, and Clawson 1997).

As "standards of behavior appropriate for a given identity," norms vary across societies as well as specific identities within a society (Finnemore and Sikkink 1998; see also Crandall, Eshleman, and O'Brien 2002). In this sense, norms also can be conceptualized as shared conventions about how different classes or types of people should act (Chudek and Henrich 2011). Although some norms operate at a conscious level, others are so deeply ingrained that they are "taken for granted," unconsciously shaping behavior and attitudes in ways individuals may not fully recognize.

The power of norms in shaping human attitudes and behaviors is widely documented (Ostrom 1998; Elster 1989). Children appear to exhibit an innate ability to notice and internalize these unstated rules of appropriate behavior across multiple cultures (Sripada and Stich 2007). Norms can generate very costly behavior, such as the formerly widespread adherence to the norm of accepting a "duel" based on perceived insults (Axelrod 1986), or failure to protect against major financial risks based on norms of personal responsibility for straying livestock that are contrary to legal rules of liability (Ellickson 1991). Norms also elicit personally costly sanctioning behavior by those witnessing noncompliance (Elster 1989; Axelrod 1986).

At the same time, norms may be ambiguous in the scope of their application, so it is not always clear which norms best apply to a particular issue (Mikhail 2011; Richerson and Henrich 2009). In this sense, multiple norms may influence the same attitude in contradictory ways. My belief in a norm that women should work in the home, for example, might give me a more negative attitude toward a specific female candidate for public office, even as my belief in a norm requiring equal treatment for all people might make my attitude toward the same candidate more positive. Thus, changing an individual's perception of which norm or norms should govern a particular situation or issue is a possible point of intervention for political actors seeking policy change.

In addition, as standards of appropriate behavior for a particular identity, norms can also vary across important cultural differences within a society. Cognitive scientists, for instance, have argued that a key differentiator of shared norms is partisanship, or one's identification as a political liberal or conservative. George Lakoff (2002) contends that conservatives are distinguished from liberals in large part by their different moral systems related to appropriate norms for parenting. Conservatives' political views, asserts Lakoff (ibid., 33), stem from their shared belief in a "Strict Father" conception of parenting that stresses personal responsibility and self-reliance, respect for authority, and obedience. Liberals, by contrast, hold political views consistent with a different set of shared beliefs about parenting, the "Nurturant Parent" model characterized by an obligation to support and protect children, the appropriateness of questioning authority, and an obligation for authority figures to explain their decision making (ibid., 34).

These divisions are also replicated in US society, at least across the major political parties that are associated with liberal versus conservative political views: Democrats and Republicans (Lakoff 2008). Variation in support for some norms according to partisanship is crucial to understanding the potential for norms to effect political change. While many norms will be appealing to both political conservatives and liberals, others may not. In seeking to appeal to norms as a justification for policy change, activists also have to consider the ways in which some norms are shared more widely than others across politically relevant identities such as party identification.

Norms vary in their ability to influence attitudes on an issue across two important dimensions—normative force and normative fit—discussed in more detail below.

Normative Force

Some norms exercise relatively weak influence over an individual's attitudes and choices, being violated with little difficulty and minimal risk of punishment. Other norms, by contrast, are violated only rarely and at the risk of serious social sanctions. In this sense, a *forceful* norm is one that substantially influences an individual's attitudes and behavior, generates costly sanctions for norm violators, and therefore is difficult for individuals to violate. Although the degree of a norm's influence over an individual's behavior will vary both within and across societies (Sripada and Stich 2007), there is little dispute that some norms are more powerful than others within a given society.

Some types of norms may be more forceful than others. For example, work in political psychology notes that moral norms (and values) are not easily traded off against other beliefs, giving them additional power over an individual's attitudes or choices (Sripada and Stich 2007; Elster 2007, 104; Baron and Spranca 1997). Similar research suggests that the use of "sacred rhetoric" or "protected" moral values against a consequentialist set of arguments may provide a political advantage, such as for those defending gun ownership in the United States on moral grounds (Marietta 2008), or that moral values are the core of many political disputes (Lakoff 2002, 2008). A related line of research documents a distinctive role for moral norms in shaping compliance with public policies through so-called affirmative motivations to follow the law (Tyler 2006; May 2005).

More specifically, moral norms of fairness and reciprocity have a distinctively strong influence over individual behavior and attitudes. Cross-cultural experiments have found that individuals regularly agree to distributions of wealth consistent with norms of fairness rather than maximizing their personal shares (e.g., Bowles and Gintis 2011; Wiessner 2009; Henrich et al. 2004; Nowak, Page, and Sigmund 2000). Although the specific norms of fairness vary (Henrich et al. 2005), behavior has been shown to deviate substantially from the predictions of models of self-interest in favor of norms of reciprocity and fairness in many societies, possibly for evolutionary reasons (Fowler and Christakis 2013; Axelrod 1986). Other work finds norms of fairness play an important role in creating support for policies ranging from health care (Lynch and Gollust 2010) to public assistance (McCall 2013; Hochschild 1981) to natural resource management (Ellickson 1991; Libecap 1989).

In addition, strongly held norms of fairness have been shown to be influential over distributional policies. Jon Elster's work on "local justice" theory is a classic example. Elster (1992, 1995) demonstrates how those making rules for distributing scarce and indivisible goods, such as school admissions or organs for transplant, are motivated by considerations of both efficiency and equity. Public opinion shapes the allocators' perceptions of equity; they desire to avoid "allocative episodes perceived as grossly unfair or wasteful" (Elster 1992, 182). Thus, actors in any allocation process must articulate their claims in terms of equity principles due to the public nature of the debate. This creates pressure in favor of allocation arguments that are consistent with a "commonsense conception of justice" that Elster (ibid., 185) describes as a "number of widely and strongly held views about justice that may not be mutually compatible." In this regard, Elster's vision of a commonsense conception of justice strongly resembles a collection of shared moral norms regarding the fair distribution of resources. A prominent norm within this commonsense conception of justice, according to Elster (ibid., 68), is the egalitarian idea of equal shares—an idea that plays a large role in the new approach to emissions allocation discussed in this book. (Notably, Elster mentions emissions allowances as a promising future application for his model even before the events in the 1990s and 2000s explored in this volume). Similar work by other scholars echoes the significance of equality and other "equity" norms in distributional policies (e.g., Young 1994; Libecap 1989).

Normative Fit

Different norms that might apply to a given decision may recommend different actions, and people often follow a particular norm with little reflection or self-awareness. This combination creates a second factor that shapes the relative influence of a norm over a specific issue or decision: perceptions of normative fit, or how persuasively a norm is thought to apply to a particular issue or decision. It is important to note that estimations of normative fit are personally subjective; they depend on an individual's judgment of a norm's applicability to a specific case as well as his or her assessment of how others will evaluate the norm's applicability.

At the same time, these assessments of fit rely on several objective factors that limit the range of norms that can be plausibly applied to an issue. Norms applied to a situation in ways that ultimately lead to logical contradictions, for example, will be more likely to be perceived as having a weak fit with the issue. New information about the facts of an issue can also lead to changes in perceptions of normative fit, such as evidence of how a situation has changed over time or varies in critical but previously overlooked ways from other cases where the norm was applied. Because of general agreement on many normative principles across a given culture, and these objective constraints on the application of those norms to a given case, evaluations of normative fit should be fairly consistent—individuals will often agree on a relatively limited group of norms that can plausibly be applied to a given issue in a given society.

Evaluations of normative fit are therefore a second important component of a norm's influence over an individual's behavior: all things being equal, norms that are judged to apply more closely to a decision will exercise greater influence than those seen to have a weaker fit. When made aware of the norm they are applying to a specific issue or behavior, individuals may become persuaded that the norm does not actually fit the situation well. In this manner, the practice of "foregrounding" a norm guiding a particular decision or action can weaken the influence of that particular norm in that situation (Petty and Cacioppo 1986), especially when the norm seems to be in conflict with other important norms held by the individual or does not seem appropriate to the decision at hand (e.g., Becker and Swim 2012).

By highlighting a norm's poor fit or conflict with other norms when applied to a particular choice, advocates can help undermine an existing practice or policy (Raymond et al. 2014). In this sense, they could be said to be challenging the "normative foundation" of the status quo. At the same time, criticism of an old norm generally must include the promotion of new or alternative norms by an organized constituency in order to create successful policy or social change (Legro 2000). In this respect, change advocates must go beyond foregrounding the existing normative frame to promote a new normative frame as an alternative conceptualization of the issue, implying a new policy design.

The strategy of normative reframing is summarized in figure 2.1. To begin with, change advocates must identify the normative frame supporting the policy status quo. Supporters may explicitly cite it, or it may be implicit in the arguments used to support the policy, but either way, those seeking policy change must identify this normative foundation. Having identified the existing normative frame, change advocates then foreground that frame—pointing out either the weakness of the norms being cited, or (more commonly) the norm or norms' poor fit with the issue in question. The goal here is to undermine the normative foundation of the existing policy, making the policy more vulnerable to change. Finally, advocates must identify and promote an alternative normative frame that recasts the issue in terms of new norms that will be perceived as fitting the issue better. If successful, this new normative frame will generate stronger support for an alternative policy design that will be difficult for defenders of the status quo to resist.

1) **Identify normative frame** supporting current policy

2) **Foreground existing normative frame**, highlighting weakness of norm(s) involved or poor fit with the issue, to destabilize policy status quo

3) **Promote new normative frame**, describing the issue in terms of alternative norm(s) with better fit for issue, implying new policy design

Figure 2.1
Key steps in normative reframing strategy.

Like other framing strategies, normative reframing relies on the potential for new issue frames to change public support for a given policy. Normative reframing strategies in some cases seek policy change by trying to change public opinion first, thereby pressuring elected officials to act. In other cases, the appeal to public opinion is implicit: change advocates make the argument directly to elected officials and their advisers about how constituents *are likely to perceive* different policy options using different normative frames in order to create pressure for policy change. Either way, the ultimate audience is the elected official or regulator with the power to change the rules, but the potential for public opinion to shift against the current policy and in favor of a new alternative is a key lever for making the policy change more likely.

In sum, it is well established that norms are a powerful influence over individual and collective decision making. Norms also vary systematically in their influence over individual attitudes and behaviors, with a substantial body of research indicating that moral norms have a distinctively strong normative force. Additional research suggests the importance of fairness norms in the design of distributional policies related to the allocation of scarce resources. Moreover, ambiguity over the applicability of different norms creates an opening to change an individual's perception of the most appropriate norm for a given issue. For these reasons, changing the norms seen as most relevant to an issue is a promising strategy for creating sudden policy changes, and especially for distributional policies such as the choice to auction emissions allowances.

The Continued Struggle to Explain Sudden Policy Change

More recent theories of the policy process have given greater attention to norms and other ideational factors, going beyond incrementalist and interest-based accounts. Despite incorporating these ideational factors, though, these theories still struggle to offer a fully satisfying theoretical explanation for specific, sudden policy changes (Cook 2010, 483; Meier 2009, 8). Instead, they tend to rely on unpredictable exogenous factors, such as natural disasters, scandals, or surprising election results, as the most common causes of major policy change. Even approaches that give greater attention to how political actors strategically promote policy change, such as PE theory, remain pessimistic about the ability of *any* theory to predict specific policy changes (Baumgartner, Jones, and Mortensen 2014). This section briefly reviews the progress made by some of these leading theories as well as the ways in which they continue to fail to adequately explain sudden policy change based on the unique power of norms.

The Advocacy Coalition Framework and Multiple Streams Theory

A leading alternative to traditional interest group theories is the advocacy coalition framework (ACF). The ACF starts from a focus on "deep core beliefs," including normative beliefs, in politics (Jenkins-Smith et al. 2014, 185) and the premise that beliefs are "the causal driver for political behavior" (Weible, Sabatier, and McQueen 2009, 122). The ACF postulates that long-standing policy subsystems are made up of two or more conflicting "advocacy coalitions," each unified by a shared set of core beliefs about the world and a particular policy domain (Jenkins-Smith et al. 2014). In this respect, the ACF sounds like a promising alternative to the interest group politics perspective on policy stability and change.

Unfortunately, ACF theorists have not yet fully developed the explanatory potential of norms in shaping policy change. For example, a substantial amount of ACF research focuses on changes in *coalitions* over time rather than on policy changes (e.g., Zafonte and Sabatier 2004; Jenkins-Smith, St. Clair, and Woods 1991). Indeed, of more than a dozen primary hypotheses offered by the ACF, only a few address policy change or stability (Jenkins-Smith et al. 2014; Weible, Sabatier, and McQueen 2009). The ACF also postulates that the stability of normative beliefs among opposing coalitions makes major policy change unlikely without a change in political power (Jenkins-Smith et al. 2014, 201–202). Of the ACF's four major pathways to policy change, two involve largely unpredictable external or internal shocks that redistribute political resources or support, the third is a more incremental process of "policy learning," and the fourth involves a variation on the interest group politics model where a "hurting stalemate" makes both sides prefer a change in the status quo (ibid., 202–203; Weible, Sabatier, and McQueen 2009). Notice that none of these pathways includes much consideration of the role of norms in facilitating sudden policy *change* or the potential to reframe an important issue in terms of a different set of normative beliefs.

A more promising alternative to the interest group perspective comes from the multiple streams theory of John Kingdon (2003). Building on Cohen, March, and Olsen's (1972) model of organizational decision making, Kingdon contends that new policies are possible only when three independent processes (or "streams") come together in a given "policy window." The streams in question are the *problem stream*, where political actors attempt to define conditions in the world as problems requiring political remedy; the *solutions stream*, where policy proposals percolate for years awaiting the right opportunity to be applied to a salient problem; and the *politics stream*, composed of events affecting the power and influence of elected political leaders. Only when the right solution finds the right problem, as it were, at the right political moment is a new policy able to be enacted.

Norms and values play an important part in setting policy agendas and predicting successful policy changes in the multiple streams model. Kingdon (2003, 109) notes that conditions are transformed into problems suitable for political action only "when we come to believe that we should do something about them." To the extent that medical care begins to be viewed as a "right," for instance, "the pressure for such government action as comprehensive national health insurance becomes much greater" (ibid., 111). For this reason, it is crucial to make proposed policy solutions congruent with shared "public values" that could also be described as moral norms. This pressure for "value acceptability" spills over into the politics stream as well, where elected officials pursue policies that match the perceived national mood (ibid., 132–137).

Similarly, the personal values of policy experts also matter; advocates promote policy ideas for years at a time based on their personal values, and policies that are incompatible with shared values or dominant norms in an expert community are unlikely to survive (ibid., 123, 132). Norms of equity and fairness are especially important: "Government officials and those around them sometimes perceive an inequity so compelling that it drives the agenda. Even if a principle of equity is not a driving force, fairness or redress of imbalance is a powerful argument used in the debates for or against proposals" (ibid., 135). This is true for both more value-laden topics like health care and more mundane issues like transportation policy, where equitable distributions of burdens and benefits are still politically vital (ibid., 94). For Kingdon, politics is fundamentally about persuasion rather than simply material interests or power (ibid., 125–127).

In many respects, Kingdon anticipates the account offered here of norm-driven policy change in the face of opposition by powerful vested interests. Unfortunately, relatively few authors have built on Kingdon's theoretical work, and in particular his emphasis on values, leaving his ideas about value acceptability and equity underdeveloped as a more general approach to explaining sudden policy change.

Policy Design Theory and Constructivist Perspectives

Other policy theories stress the importance of norms and ideas from a constructivist perspective. These approaches emphasize the political significance of shared, "socially constructed" understandings of reality as opposed to allegedly objective assessments of specific material interests (Onuf 1998). For example, policy design theory describes how society's perception of a social group determines the kind of policy used to meet that group's needs (Schneider and Sidney 2009; Schneider and Ingram 1993). This is a reciprocal relationship, in that a policy design may also reinforce a group's social construction, such as a policy that treats drug addicts as "deviants" and thereby reinforces negative public perceptions of that group, making even more punitive future policies likely (Schneider, Ingram, and DeLeon 2014; Mettler and Soss 2004). On this basis, policy design theory offers a few general empirical propositions, including that policies for advantaged groups will be more generous and easier to adopt politically than policies addressing the needs of individuals with little political power and negative social constructions (Schneider and Ingram 1993; Schneider, Ingram, and DeLeon 2014, 113–115).

Policy design theory explicitly emphasizes "the integration of normative and empirical analysis" in policy research (Schneider and Sidney 2009, 112), again consistent with the importance of normative beliefs stressed by this chapter. By *integration*, policy design theorists appear to mean two things: evaluating existing and proposed policies in terms of various normative ideals, such as justice, equality, and democratic freedom; and including normative "constructions" of target populations in explanations of why certain policies are more or less likely to arise. Although both are crucial, leading statements of the approach stress the first goal in particular (Schneider and Sidney 2009; Schneider and Ingram 1997). This makes policy design theory less of an empirical theory of the policy process than the others reviewed in this section, or at least a more incomplete theory of policy stability and change than is required to explain or predict a sudden policy shift, such as the adoption of allowance auctions in RGGI and subsequent emissions trading policies.

Other accounts of policy change in the constructivist tradition have focused on explaining differences in national economic and social welfare policies in a more comparative context (e.g., Berman 2001; Thelen 2004). Studies in this area have identified beliefs about the appropriate role of government in the economy (Berman 1998; Blyth 2002), for instance, or norms regarding traditional family roles (Boling 2015; Fleckenstein 2011) as key explanatory factors for divergent economic and social welfare policies among nations. As with policy design and multiple streams theories, the focus is on alternatives to explanations based primarily on economic self-interest and political power.

This literature on comparative social policies also highlights the role of noneconomic factors, including norms, in driving policy change. But even in this area of scholarship, explanations are frequently grounded in the idea of path dependence, in which key decisions taken at a particular "critical juncture" in policy development lock in a policy choice for years or decades to come (Pierson 2000; Thelen 2004). Although some work in this tradition has concentrated more on explaining major policy change (e.g., Streeck and Thelen 2005), the historical institutional perspective that underlays much of this work continues to struggle with explaining or predicting the occurrence of these critical junctures rather than the path dependence that follows (Schmidt 2008; Peters, Pierre, and King 2005; Berman 2001).

A final constructivist perspective on norm-driven policy change emerges from work in international relations on norm adoption and diffusion (Sikkink 2011; Risse 2000; Finnemore and Sikkink 1998). As with the other perspectives described above, the focus is on how national leaders "construct" their world instead of taking their material circumstances as self-evident. In this tradition, nations are thought to choose policies in part according to a logic of appropriateness driven by international norms of civilized behavior. That focus on norms allows nongovernmental actors to function as "norm entrepreneurs," pressuring nations to adopt new policies that adhere to new or emerging international norms regarding human rights or other issues (Friedman, Hochstetler, and Clark 2005; Weyland 2005). Here activists use norms as tools to promote policy change through social pressure along with attempts to highlight a nation's positive or negative identity in the global community (Sikkink 2011; Finnemore and Sikkink 1998). As with work in historical institutionalism, this line of research also helps explain how pressure to comply with accepted norms can lead decision makers to adopt new policies. At the same time, the primary interest of this literature is the diffusion of new norms rather than reframing an issue in terms of an alternative norm already accepted in a particular society, thereby distinguishing it from the theory developed here.

PE Theory

PE theory draws on a metaphor from evolutionary biology to depict a policy process dominated by incremental change with occasional "punctuations" resulting in substantial policy shifts (Jones and Baumgartner 2012). As with the other theories described above, PE theorists seek to move beyond the interest group politics understanding of policy stability through greater attention to the role of ideas, including norms, in the policy process (Baumgartner and Jones 1993, 22). PE theory also articulates a mechanism by which interest groups can use new policy images to promote policy change that is different from other theories reviewed and much closer to the idea of normative reframing. As such, it is the most helpful theory for building a satisfying explanation for the sudden adoption of auctioning in GHG emissions trading policies and explaining sudden policy change in general.

A core concept of PE theory is the limited resource of attention in the policy system (Jones and Baumgartner 2005, 2012). Because political actors can attend to only a handful of issues at a time, most policy making happens at the subgovernment level, where a small number of actors with significant interests and shared values maintain their preferred status quo (Baumgartner and Jones 1993, 19). At the same time, issues occasionally jump to the "macropolitical" level of broader attention, where more substantial changes are possible. Scarcity of attention means that only a few issues will move from micro- to macropolitics at a time, leaving most of the policy process controlled by smaller "policy monopolies" (similar to the iron triangles of interest group theories) opposed to significant change.

PE theory in this respect combines a more traditional explanation of policy stability at the level of micropolitics with a new account of policy change at the macropolitical level. Thus, the vital question becomes how do policies move from the micro- to the macropolitical agenda. For PE theorists, *policy images* are vital to this process. Such images are "public understandings of policy problems" (ibid., 25) or "the manner in which a policy is characterized or understood" (Baumgartner, Jones, and Mortensen 2014, 83). Although many policies have a largely uncontested image, advocates can promote a new policy image in order to move an issue to the macropolitical agenda—a nonlinear punctuation of an otherwise incremental process (ibid., 69). In some cases, new images will invoke new norms, such as when poverty is recast as a "structural" public problem rather than an issue of "personal misfortune" (Baumgartner and Jones 1993, 27–8, citing Majone 1989). Once a new image and resulting policy change is established, the policy returns to the subgovernment level described by interest group politics theories, unlikely to change again until another punctuation.

Policy images closely resemble issue frames, making normative reframing closely related to PE theory. Yet PE theory does not focus on the unique power of norms to drive policy change, stopping short of trying to explain *why* certain policy images succeed or fail as causes of policy change. In some cases, researchers working in the PE tradition even seem relatively unconcerned about the relationship of new policy images to "facts on the ground"—coming close to suggesting that one policy image is as plausible for a given issue as another (e.g., Pralle 2006, 157). More recent efforts to explain and predict specific policy changes using PE theory have centered instead on variations in the degree of political attention being focused on an issue as opposed to the *causes* of those variations in attention (Hegelich, Fraune, and Knollmann 2015). In short, the reasons why specific policy images succeed or fail at increasing political attention as well as sparking policy change remain substantially unexplored in PE theory. PE theory would benefit from greater attention to both the distinctive power of policy frames (or images) invoking strong social *norms* as well as the idea of normative fit—that is, how *persuasive* the application of a new (normative) frame (or policy image) is to the accepted facts of an issue in the minds of the frame receivers.

Perhaps because of this lack of consideration of how well different policy images fit a given issue, PE theory is pessimistic about predicting specific policy changes. Although the theory predicts the general distribution of policy changes in a given issue area over time, it "will not help us make point-specific predictions for particular policy issues" (Baumgartner, Jones, and Mortensen 2014, 91). Building on the insights of PE theory, this book will argue that although "point-specific" predictions of large policy changes at a future moment in time are always unlikely at some degree of precision, more attention to the distinctive influence of normative frames along with the idea of normative fit should permit better explanations of *why a particular policy change occurred* in the past as well as *which policies are more or less likely to face sudden policy change* in the future. The next section reviews how scholars and advocates might use the theory of normative reframing to better explain sudden policy changes, and to assess which policies are more "ripe for change" due to their relatively weak normative foundations.

Normative Reframing: An Important Mechanism of Sudden Policy Change

As noted above, the concept of normative reframing connects the idea of new policy images from PE theory to the more general notion of framing in political psychology by conceptualizing policy images as alternative frames for a given issue. The theoretical argument of this book is that normative reframing offers a unique ability to explain and predict policy changes in the face of opposition from powerful interests. To better understand the idea of normative reframing, it is important to first review some core ideas about framing, attitudes, and political change. This section will then explain how the ideas of normative force and fit can help identify stronger versus weaker normative frames for a given issue as part of a normative reframing strategy for policy change.

Framing in Political Communication

According to a leading scholar of political communication, frames "select some aspects of a perceived reality and make them more salient ... in such a way as to promote a particular problem definition, causal interpretation, moral evaluation, and/or treatment recommendation" (Entman 1993, 52). A substantial body of evidence has shown that different frames can generate different levels of support or opposition for the same policy, changing public attitudes in a process known as a "framing effect." Framing effects have been documented across a wide range of policy issues, from social policies (Bali 2009; Craemer 2009; Brewer 2003) to government spending (Jacoby 2000) to energy and environment issues (Lockwood 2011; Nisbet 2009).

Some frames are logically equivalent ways of stating the same idea, such as noting that a policy will create 90 percent employment versus 10 percent unemployment. Many framing studies have documented how even logically equivalent frames can create different attitudes toward the same policy (in general, see Tversky and Kahneman 1981), causing some observers to worry about the apparent irrationality of public attitudes (e.g., Sunstein 2002). A more common way of thinking about frames in politics, however, is as competing constructions of an issue that "spell out the essence of the problem" in different ways, thereby recommending different government actions (Nelson and Kinder 1996, 1057). These "issue frames" are not logically equivalent; they highlight different aspects of a particular issue in an attempt to sway public or elite attitudes (Druckman 2004).

Issue frames can change attitudes in two major ways: either by giving greater emphasis to certain "accessible" beliefs already being used to form an attitude, or by making alternative beliefs "available" for the frame recipient to apply to the issue for the first time (Chong and Druckman 2007). Many frames are thought to increase the "weighting" that individuals give to a particular belief they were already using to help form an attitude about that issue (Nelson and Oxley 1999). For example, describing a Ku Klux Klan rally as a "freedom of speech issue" rather than an issue of "public order" leads to higher levels of support for allowing the rally by increasing the weight given to belief in the importance of free speech in forming an attitude on the issue (Nelson, Clawson, and Oxley 1997).

In making a new belief available, by contrast, a frame facilitates the application of an existing belief to a new issue. An "innocence frame" for

capital punishment, suggesting it is wrong to risk putting even one inno-
cent person to death, for instance, has been shown to have a greater ability
to change attitudes toward the death penalty than more familiar frames
emphasizing the immorality of taking an individual's life (Baumgartner,
De Boef, and Boydstun 2008). In such cases reframing makes a new belief
available, thereby creating a more effective "conflict displacing" frame that
focuses on an aspect of the issue that both sides might support instead of
emphasizing established points of disagreement (Dardis et al. 2008).

It is important in either case to note that frames are not changing individuals' beliefs but rather helping them to view particular beliefs as more
or less important to their evaluation of the issue. In the case of making a
new belief available (or dramatically reweighting the relevance of existing
beliefs in forming an attitude), successful frames get individuals to see the
issue as "a case of this" versus "a case of that," changing their attitudes
without forcing them to change any of their deeply held beliefs. This is the
strategy that underlies the idea of normative reframing, where a new frame
recasts an issue as subject to a different norm or norms, thereby generating new attitudes without challenging existing normative beliefs that are
resistant to change.

Creating the Strongest Possible Normative Frame: Force and Fit

All things being equal, a frame's "strength," or its power to change attitudes, can be attributed to the strength of the beliefs being invoked by the
frame as well as the frame's ability to appeal to the audience (Chong and
Druckman 2007, 111). In the case of normative frames, these two qualities
translate directly to the ideas of normative force and normative fit described
earlier in this chapter.

Existing work on framing indicates that reframing an issue in terms of
a *more strongly held belief* will increase the frame's effect on an individual's
attitude toward a particular issue. In other words, frames that invoke more
"central" beliefs in a society (Benford and Snow 2000, 621) or "longstand-
ing cultural values" are likely to be more influential on public attitudes
(Chong and Druckman 2007, 112; see also Chong 2000). In this sense, the
most effective normative frames will present an issue in terms of norms
with the greatest normative force possible.

At the same time, a frame must "appeal" to its audience; it must be persuasive to be effective. This idea of a frame's persuasiveness assumes an

"active receiver" model of framing, where recipients of a framing message deliberate over how convincing they find the frame's message to be (Brewer 2001; Sniderman and Theriault 2004). As active receivers, frame recipients moderate the effect that a frame has on their attitudes by considering how applicable the frame is to a given issue or decision as well as how consistent the frame is with their personal values and norms (in general, see Chong and Druckman 2007, 110–111). It is this question of what makes a frame more appealing or persuasive that is at the heart of the idea of normative fit presented in this chapter.

Work on framing by social movement scholars provides an excellent parallel to the ideas of normative force and fit explored here. For example, Robert Benford and David Snow (2000) describe how frames that "resonate" with the public are more effective at influencing behavior, where resonance is conceptualized as a function of the frame's *salience* and *credibility*. A frame's salience on this account is defined partly by "how essential the beliefs, values and ideas associated with the ... frame are to the lives of the targets" as well as the frame's consistency with the recipients' lived experiences and dominant cultural narratives and assumptions (ibid., 621). These qualities closely parallel the factors determining a norm's force discussed earlier in the chapter.

On the other hand, the idea of a frame's credibility is directly related to the notion of normative fit. Benford and Snow (ibid.) explain how credible frames encompass a set of beliefs and claims that avoid self-contradiction, and appear to be consistent with the actions supported by the advocates promoting the frame. Even more important is the idea of the frame's "empirical credibility"—that is, the "apparent fit between the framings and events in the world" (ibid., 620). In this regard, effective frames are consistent with the audience's perceptions of the facts of a given situation.

This evaluation will vary, of course—messages that are highly implausible to the general public may be credible to members of a cult, say, or other small groups, due to different beliefs about the facts of a situation. Even members of different political parties in countries such as the United States, for example, tend to hold different empirical beliefs about the causes and severity of climate change (McCright and Dunlap 2011; Kahan 2010). Given this disagreement about the science of climate change, some have argued that frames promoting new climate policies on the basis of other, more widely shared beliefs such as the importance of generating broad economic benefits for citizens will be more promising politically (Skocpol 2013). The new normative frame promoting the public benefits of auctioning allowances is consistent with this approach by focusing on tangible economic benefits and protections for the public from a new climate policy rather than the projected risks of climate change.

In this respect, leading social movement theorists anticipate the model offered here by describing how other researchers fail to "take seriously the constraints that 'culture out there' imposes on ... framing activity" (Benford and Snow 2000, 622). In the current discussion, a critical aspect of "culture out there" is the relatively fixed nature of many powerful social norms and relatively limited applicability of specific norms to various issues. Social norms are also somewhat ambiguous in their scope and subject to change over time, which suggests that the set of potentially relevant norms for a given issue may vary over longer time frames, even as it remains relatively fixed at a given point in time (ibid.; see also Raymond et al. 2014).

Finally, it is important to recognize that the motivations of policy change advocates in selecting a new normative frame can be both strategic and genuine. Given their desire for policy change, advocates are unlikely to favor the norms currently applied to the policy status quo. In this sense, they are looking for alternative standards of appropriateness that they genuinely believe are better applied to the situation. The choice of a new normative frame is at the same time strategic; all things being equal, change advocates will choose an alternative frame they expect to be most persuasive *to the relevant audience* in order to maximize the new frame's expected normative fit and therefore its overall strength. Environmental advocates, for instance, might have to choose a normative frame they believe will be more effective with the public (or a particular group such as Republicans or Democrats) over one they personally prefer, such as promoting renewable energy in terms of improving national security versus protecting the environment. The choice of a normative frame, then, combines genuine and strategic motivations.

Summing Up: Generating a Strong Normative Frame

All things being equal, the frame that invokes the norm with the *greatest force* and *best fit* for the issue will have the strongest influence over attitudes in a given situation. An alternative normative frame may be weaker due to its invoking a less forceful norm or the weaker perceived fit between the norm and the issue. This conceptual summary of the strength of a normative frame as a product of the force of the underlying norm coupled with the fit between the norm and the issue itself is presented in figure 2.2.

Thus, policy advocates face a dilemma in choosing a new normative frame for an issue: how to trade off normative force versus fit. Normally, advocates would be expected to choose issue frames invoking norms that are deeply and widely held—that is, as "forceful" as possible. One sees evidence of this, for example, in efforts to promote many new policies as human rights issues because norms against violating most human rights are so deeply held. At the same time, some issues are more convincingly portrayed as human rights issues than others for a given society, making that frame's normative *fit*, and corresponding ability to influence attitudes and policy choice, vary significantly across issues.

In this sense, one can imagine an advocate of policy change weighing the trade-offs between wanting to invoke as forceful a norm as possible and the relative difficulty of persuading others that the norm fits the issue in question. A weaker norm may be more politically effective in some cases because it fits the issue more persuasively in the perception of decision makers or the public. Alternatively, when norms with relatively equal force are used in competing frames, the degree of fit with the issue will be decisive in shaping the relative influence of the alternative frames. This is exactly the situation that arose in the successful promotion of auctions in RGGI, where the weak fit of a forceful norm with the allocation context created a policy that was ripe for change based on new normative framing.

Normative Reframing and Allowance Auctions

Having outlined the general theory of normative reframing as a strategy for policy change, it is now possible to introduce the application of that model to the decision to auction emissions allowances. As noted in chapter 1, the model suggests that environmental advocates recognized and attacked what they viewed as a weak frame supporting the policy status quo of

$$\begin{array}{c} Normative\ frame \\ strength \end{array} = \begin{array}{c} Normative \\ force \end{array} \times \begin{array}{c} Normative \\ fit \end{array}$$

Figure 2.2
Simple model of determinants of a normative frame's strength.

grandfathering allowances. These advocates then chose to reframe allocation policy in terms of a new frame relying on two alternative norms: that
polluters should pay for their use of a public resource, and that the benefits
from those public resources should be distributed in a *fair and egalitarian
manner*. This public benefit frame was accompanied by an associated policy
design that auctioned emissions allowances *and* dedicated the revenue to
programs benefiting most citizens. In this respect, normative reframing
requires a careful match between the new frame being applied to the issue
and the alternative policy design being promoted. In the case of the shift
to auctioning allowances, the combination of the new public benefit frame
and policy design can be referred to collectively, as mentioned earlier, as the
public benefit model.

Advocates reframed the allocation issue in terms of these alternative
norms because they believed the norms supported the idea of auctioning
allowances, and that most policy elites and the majority of the public would
see them as having a better fit with the issue of allocating emissions rights.
Although it was not the only factor that led to this decision to make emitters pay for their emissions, a review of the RGGI case along with its important precedents and successors will suggest that this reframing appears to
have played a crucial role in making this major policy change possible. This
section presents the normative reframing process in greater detail, describing how the better fit of the alternative frame supporting auctions helped
spur the policy change to auctions for the first time in RGGI.

The Weaker Normative Frame Justifying Free Allocation

Figure 2.3 summarizes the normative framing facing those seeking a change
in allowance allocation. The existing frame being used to support the policy
of free allocation of allowances to current emitters applied a norm of private entitlement to a resource based on beneficial prior use, consistent with
the property theories of John Locke. This private entitlement frame used
the Lockean norm to justify the free allocation of rights to use the atmosphere to current emitters, based on their prior use of that resource. Figure
2.3 clarifies that although the norm rewarding beneficial prior use with
ownership is quite powerful (at least in the United States), it demonstrates
a relatively weak fit with the issue of allocating emissions rights because it
requires defining the emission of pollution as a beneficial use comparable
to growing crops or building homes. The result of this particular normative

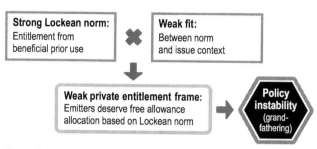

Figure 2.3
Weak private entitlement normative frame for grandfathering allowances.

framing, therefore, was a weak normative foundation for the policy of grandfathering, putting the policy at relatively greater risk of change.

There is little question that the idea of beneficial prior use is a common and influential distributional norm. It is perhaps best represented by the writings of Locke ([1690] 1994), who conceived of labor as the ultimate justification for the ownership of resources, with certain qualifications. This Lockean ideal was a guiding principle in the creation of American society, and remains influential in the modern era (Hartz 1955; Nedelsky 1990). The norm has shaped policies to distribute numerous environmental and natural resources (Libecap 1989; Raymond 2003), including but not limited to agricultural land, mineral rights, water rights, range forage, and fisheries. It also appears in many more informal settings, such as the common law doctrine of adverse possession or (more vividly) in the widespread practice in several major American cities of individuals asserting "ownership" of parking spots on public streets after shoveling their cars out following a snowstorm (Buying Property with a Shovel 2010; figure 2.4).

Although the Lockean idea of rewarding beneficial labor is a powerful one, it is normally applied to actions such as farming, mining, or otherwise directly producing valuable goods. In the case of cap-and-trade policies, chapter 3 will illustrate how policy makers extended the Lockean norm without much reflection to a different type of resource use: emitting harmful pollutants into the atmosphere. On greater scrutiny, this application of the norm became difficult to defend because it relies on applying a norm rewarding beneficial use to what is widely regarded as a harmful activity. Although it is possible to argue that polluting is merely an unavoidable side effect of other beneficial activities, such as generating electricity, even the quickest review of the varying degrees of pollution from different energy

Figure 2.4
The Lockean norm in action (photo credit: Lauren Zawilenski).

sources and power plants makes that assertion difficult to maintain. Simply put, polluting is a harmful and significantly avoidable activity, making the application of the Lockean norm to this case problematic for even strong supporters of this norm.

Thus, figure 2.3 indicates how the private entitlement frame supporting the status quo policy of free allocation was relatively weak, making the policy vulnerable to sudden change through normative reframing. To do this, advocates first had to recognize and highlight the weakness of the normative fit between the widely respected Lockean norm and the issue of allocating emissions rights. By pointing out the unpersuasive nature of a frame justifying free allocation to polluters based on a norm rewarding beneficial prior use, environmental advocates destabilized a policy that favored powerful interests and therefore was not expected to change.

The Stronger Normative Frame Justifying Allowance Auctions
Absent a better alternative, criticism of an existing normative frame is unlikely by itself to generate policy change. Thus, environmental advocates

also had to identify and promote alternative norms related to allocation through a different frame. Responding to this challenge, RGGI auction supporters offered and promoted the new public benefit frame describing emissions allowances as *public* assets that should be distributed in a manner that helps most citizens in a tangible way. Figure 2.5 illustrates how this reframing brought two influential alternative norms to bear on the allocation issue: the widely recognized *polluter pays norm* suggesting that firms should bear the costs of their emissions, and an *egalitarian norm* favoring broadly distributed and relatively equal benefits from any private use of a public resource. Like the Lockean norm favoring prior use, the polluter pays and egalitarian norms have strong normative force among most of the public, at least in the United States and Europe. Unlike the Lockean norm, however, figure 2.5 indicates that these alternative norms were perceived to fit the allocation issue context much better than the Lockean entitlement based on beneficial prior use.

The polluter pays norm is one of the oldest and most widely supported in the field of environmental policy. Based on more general norms of personal responsibility, the idea first entered environmental policy discussions in the early 1970s in a report by the Environmental Directorate of the Organization of Economic Cooperation and Development (Pearce 2002), although the concept of holding polluters responsible for the harm caused by their actions dates back centuries in common law (Gaines 1991; Coase 1960). The principle has influenced many formal environmental policies in the United States, the European Union, and the world (Gaines 1991; Nash 2000, 467), and remains widely supported by the public (Swedish Environmental Protection Agency 2010; Lange, Vogt, and Ziegler 2007).

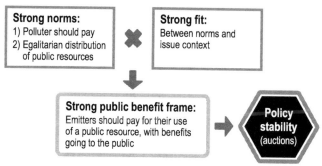

Figure 2.5
Strong public benefit normative frame for auctioning allowances.

Egalitarian norms are also common in many societies and distributional contexts (Sripada and Stich 2007; Young 1994; Elster 1992). As mentioned earlier in the chapter, individuals given unilateral power to divide a sum of money in so-called dictator games tend to select a relatively equal distribution rather than taking most or all of the money for themselves. Similarly, in a variant of these types of experiments known as "ultimatum" games (in which the second person can reject or accept the distribution proposed by the first player), many players reject extremely unequal offers, creating a result where both players receive no money at all. Although the specific results vary, the general trend toward more equal distributions is consistent across many cultures, especially in the United States and Europe (Henrich et al. 2004), and across members of conservative and liberal political parties (e.g., Dawes et al. 2012). These economically irrational actions indicate the power of norms recommending equal distributions, especially when there is no suggestion that one subject or the other has "earned" a greater share of the funds to be divided (Franco-Watkins, Edwards, and Acuff 2013).

An egalitarian norm is also influential in many policy settings, ranging from human rights declarations to more mundane examples of distributive policies. In the case of human rights, policy advocates draw on norms shared by many societies that all people are created equal and hence entitled to the same basic rights. As noted above, these norms can be effective in motivating governments to change their policies in order to avoid being perceived as a nation failing to live up to generally accepted standards of civilized behavior. In addition, it was noted earlier in the chapter how local distributional policies often follow egalitarian norms as part of a "commonsense conception of justice," giving everyone an equal share of a resource, or at least equal opportunity for access (Elster 1992, 1995). More generally, negotiations over distributions of natural resources frequently reflect the influence of this norm of equality (e.g., Libecap 1989; Zerbe and Anderson 2001; Young 1994), as do rhetoric and policy designs related to environmental protections of air and water quality based on the idea that all citizens have the right to clean air and water.

Some may argue that polluter pays and egalitarian norms are less popular than the Lockean norm favoring prior use, especially among those who identify with conservative political parties such as the Republican party in the United States (e.g., Lakoff 2002, 170). Although evidence on this point is scarce, there are good reasons to suspect that both norms are consistent with key principles of diverse political ideologies, making them especially useful for partisan climate policy conflicts. The polluter pays norm, for example, is grounded in the importance of taking personal responsibility for one's actions—a fundamental part of the conservative set of moral values outlined by Lakoff (ibid.). Furthermore, many prominent Republican policy proposals draw on strict versions of equality as a key norm, including opposition to affirmative action and support for a flat tax. Opposition among Republicans to the redistribution of wealth is clearly greater than among Democrats (Esarey, Salmon, and Barrilleaux 2012), but the notion of distributing relatively equal shares of value among all citizens is more consistent with Republican orthodoxy than the idea of using auction revenue to reduce wealth inequality.

More generally, studies suggest that similar fairness norms are influential across party lines in both lab experiments (Dawes et al. 2012) and actual policy discussions (Kingdon 2003). In the case of the push for auctions of emissions rights, this book shows that conservative interests typically did not challenge the polluter pays and egalitarian norms directly but instead relied on alternative arguments that lacked the same persuasive power. Thus, there are good reasons to think that these two norms are influential among conservatives as well as liberals when applied to polluting and the use of the atmospheric commons.

Figure 2.6 summarizes this full model of norm-driven policy change proposed to explain the adoption of auctions. First, policy advocates foregrounded the weakness of the private entitlement frame supporting grandfathering based on its poor fit with the issue. Policy advocates then offered an alternative, public benefit frame applying two different strong norms to the issue, suggesting that polluters should pay for their emissions and that value should be widely distributed among the public. This new framing was more politically successful because of the greater perceived fit of these new norms to the issue. The new framing had the additional advantage of focusing the conflict on new facts about protecting consumers from energy price increases rather than fears about threats from climate change that are less widely agreed on across partisan identities.

As will be described in chapters 3 and 4, advocates in the RGGI case made the crucial assessment that elected officials would better support the application of a polluter pays norm to a case of pollution regulation when it was combined with egalitarian norms demanding broad public benefits

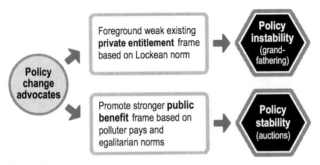

Figure 2.6
Full conceptual model of norm-driven cap-and-trade policy change.

from any private use of that resource. A careful review of the RGGI process from 2003 to 2007 (especially in comparison to its immediate emissions trading policy predecessors) suggests that these advocates were correct, and the better perceived fit of the new norms with the allocation issue was a vital part of the success of the effort to adopt allowance auctions. Moreover, figure 2.6 indicates that this stronger normative frame should offer greater policy stability in the future for cap and trade with auction or other policies based on the public benefit frame—an idea that will be explored further in chapter 5.

Thus, the improved fit of the new public benefit frame helped make a policy forcing emitters to pay for their allowances politically viable, contrary to the expectations of traditional theories of interest group politics. This is not to claim that normative reframing was the only way advocates promoted auctioning or was the only reason why the RGGI's designers took the surprising step of auctioning emissions rights. It is to suggest, however, that the new public benefit frame and associated policy design directing auction revenue to programs offering tangible benefits to many citizens was a critical factor in distinguishing the RGGI policy process from similar cases where auctioning was not taken as seriously. In particular, the addition of the egalitarian norm to generate an alternative public benefit frame and policy design going beyond the basic idea that the polluter should pay appears to have been critical to the political success of auction proposals in RGGI.

More generally, the model of normative reframing as a pathway to sudden policy change presented here does not imply that such reframing will be effective in every case. Policies supported by relatively weak normative frames are likely to remain stable until being identified and foregrounded by motivated advocates for change. Reframing efforts may also fail, especially if they are unable to find a strong alternative norm that is perceived to have a better fit with the issue. In this respect, some policies supported by weak normative frames may remain unchanged for long periods of time.

Additionally, even successful normative reframing does not guarantee the ability to change the status quo. Sudden policy change in the face of opposition by powerful interest groups remains difficult to achieve for many of the reasons described by the policy theories summarized in this chapter. Partisan opposition in particular is a growing obstacle to policy change—one that has been especially critical in the case of climate change policy since 2008. But normative reframing offers a distinctive strategy that in many cases increases the odds of overcoming opposition to change by vested material interests or partisan opponents, and adds important details to the idea from PE theory of using new policy images to create policy change. In this respect, the idea of normative reframing improves not only our understanding of the RGGI case and the broader movement toward a public benefit model in climate policy but also the potential for sudden policy change for many other policies justified by weak normative frames.

3 Economics Is Not Enough: The "Old Model" of Cap and Trade

To fully understand the significance of the choice to auction allowances for the first time in RGGI, one must look at the history of efforts to use emissions trading to combat pollution. In the preceding decades, emissions trading programs were adopted reluctantly and gave allowances to current emitters at no cost, contrary to the recommendations of many economists. This long-standing approach to emissions trading can be referred to as the "old model" of cap and trade. This chapter describes how that model largely controlled US emissions trading policy from its origins in the 1970s through the 1990s, including the enactment of the most famous and ambitious cap-and-trade program of the twentieth century: the US Acid Rain Program (ARP), passed as part of the Clean Air Act (CAA) Amendments of 1990.

Despite the endurance of the old model in this period, the 1990s and early 2000s included important seeds of change for the sudden switch to auctions in RGGI discussed in chapter 4. This chapter also reviews these new factors, which include greater attention to allocation rules by political actors, new precedents for auctioning rights to or taxing the use of other public resources, as well as new political and economic pressures from electricity deregulation. In addition, this period saw the initial emergence of new polluter pays and public ownership normative frames in the context of emissions allowance allocation.

At the same time, the chapter documents how those initial changes failed to overturn the pattern of giving emissions allowances away at no cost, even in the initial phase of the EU ETS from 1998 to 2005, or in the NO_x budget emissions trading program from 1994 to 2005 that engaged many of the same agencies and states involved in the RGGI design. Reviewing the failed attempts to promote auctions in these two close precedents to

RGGI helps sharpen the analysis in chapter 4 because it permits a comparison to similar cases where auctions were considered and largely rejected.

The Old Model of Cap-and-Trade Policy Design

Early work on emissions trading and market-based pollution policies is grounded in the economic theory of externalities. Economists have long noted that firms will ignore the social costs of making their products if those costs are not captured in the market. The price of a product, in other words, will not include the cost of environmental damage from its production unless the firm is forced to bear those costs by regulation. For this reason, a firm will produce more of its product (and create more environmental damage) than is socially optimal by "externalizing" the environmental costs of producing its goods and services onto society as a whole.

Starting with the work of Pigou (1920) in the early twentieth century, economists have proposed solutions to this problem that emphasize "internalizing" the externalities by making polluters pay for the damage caused by their emissions. According to Pigou, governments should estimate the damage created by a firm's pollution or other "external" costs of production, and impose an equivalent tax on the pollution source, thereby forcing the firm to incorporate this social cost in its production decisions. For several decades, this notion of a Pigouvian tax was the dominant recommendation of economists to the externality problem for pollution, consistent with the polluter pays norm introduced in chapter 2.

An alternative to Pigouvian taxes emerged in the 1950s and 1960s: the idea of creating tradable rights for the use of natural resources. According to Coase (1960), the Pigouvian approach could actually decrease total social well-being by fixing liability with the polluting party and created moral hazard problems by discouraging those being injured from taking low-cost steps to avoid the damage. Illustrating his point with examples of conflicting land uses from English common law, Coase described how each person disturbed by an externality of production could just as logically be said to be harming the producer of the externality. In other words, Coase rejected the polluter pays norm in advocating a more reciprocal notion of harm in which he claimed courts generally should (and did) try to assign the "rights" where they would create the most social good.

Because of this reciprocal nature of harm, however, Coase also argued for creating *transferable* rights whenever practical to allow the parties to decide where the rights would be most valuable rather than having the courts or government make that assessment. The now-famous "Coase theorem" demonstrated that voluntary negotiations would end up assigning the rights to the same party regardless of who received them initially from the government, as long as the rights were transferable and there were minimal transaction costs. By asserting the importance of tradable rights to minimizing social costs, Coase created the intellectual foundation for the idea of emissions trading policies as an alternative to an emissions tax.

Although examples of policies creating transferable private rights to limit use of natural resources date back to the nineteenth century (Raymond 2003; Macinko and Raymond 2001), the first application of the Coasean idea to air or water pollution came in the mid-1960s when Thomas Crocker (1966, 80–81) proposed creating a market in air pollution rights by having both emitters and "recipients" of pollution bid on rights to "the use of the air's waste-disposal dimension" in a manner similar to modern auctions for emissions rights under cap and trade.

A more complete vision of what is now referred to as a cap-and-trade policy for pollution first appeared in John H. Dales's (1968) book *Pollution, Property, and Prices*. Dales proposed setting a fixed limit on the total pollution from all sources into a body of water. To enforce this pollution "cap," Dales described how each source must hold an "allowance" for every unit of pollution it emits, surrendering those allowances to regulators annually based on its total emissions. Under this scheme, the cap maintains the environmental integrity of the policy even in the face of industrial expansion because new sources may enter the market only by acquiring allowances from existing emitters. Polluters may also trade these allowances according to their ability to reduce emissions more or less cheaply. In this respect, allowances are a revocable property right like a license: transferable rights to use a scarce resource that may be withdrawn by the government without compensation (Cole 2002, chap. 3; Raymond 2003, chap. 2).

Advocates promote the cap-and-trade idea based on two advantages. First, the policy is *more environmentally effective* by limiting the total amount of pollution rather than the rate of pollution. Traditional pollution control regulations take a rate-based approach, limiting the amount of pollution from a source per unit of energy produced or consumed, or taxing each unit of pollution to discourage emissions. The CAA of 1970, for instance, required power plants to reduce their pollution rate per unit of fuel burned, but did not limit power production or fuel consumed. This approach allowed expanded energy production to offset some emissions reductions from better pollution controls. With a fixed cap, new sources may not increase the total emissions under the cap, but instead must acquire allowances in some manner from existing emitters.

The second asserted advantage of cap and trade is that it is *more economically efficient* because making these rights tradable minimizes the total cost of compliance. Traditional rate-based regulations often require diverse emissions sources to meet the same emissions standard or install the same equipment, regardless of variations in the cost of reducing emissions across facilities. By contrast, cap and trade creates a market in pollution rights so that emission reductions can be made where they are least expensive. Rather than reducing their own emissions at high costs, some emitters might buy allowances from firms with lower pollution control costs, thereby moving emissions reductions to where they are cheapest. By equalizing the marginal costs of pollution abatement in this manner, the cap-and-trade approach can reduce emissions at the lowest-possible social cost (Montgomery 1972).

In this way, early economic discussions of pollution control all recommended making polluters pay for their emissions in some manner as a method of internalizing these external costs. Both Dales and Crocker, for instance, suggested that firms should buy their emissions allowances rather than receiving them for free, while others stressed a Pigouvian approach using pollution taxes and fees (e.g., Baumol and Oates 1971; Kneese and Schultze 1975). Twenty years after Dales, prominent economists advocating for cap and trade still argued for auctions as the best allocation option, even as they recognized the political challenges of making firms pay for emissions allowances (Hahn 1989; Dudek and Palmisano 1988).

It is important to note that these economic arguments did not stress the polluter pays norm but rather the general efficiency gains of using an auction (e.g., Tripp and Dudek 1989; Hahn 1983; Eheart, Brill, and Lyon 1983) to make sure the rights went to those who valued them the most. Indeed, some economists recommended a "zero-revenue" auction, in which all auction revenue was returned to the affected firms, thereby making the allocation more efficient without imposing a new cost on polluters (Hahn 1983; Tietenberg 1985, 102). Others, though, suggested the potential value of auction revenue for the government (e.g., Ackerman and Stewart 1985)—an advantage also cited by Dales back in 1968 that would finally be recognized in the RGGI case.

The first cap-and-trade policies developed gradually with changes to the CAA in the mid-1970s, and proceeded in ways that deviated substantially from the theoretical discussions happening in the economics literature (Pérez Henríquez 2013). EPA administrators first considered the trading of pollution credits due to CAA limits on the creation of new pollution sources in areas failing to meet the federal clean air standards. Confronted with the prospect of having to forbid new economic development in many major metropolitan areas that were in "nonattainment" with federal standards, the EPA faced tremendous pressure to create more flexibility in their rules.

The result was the first set of federal emissions trading provisions, allowing for the exchange of credits to pollute in nonattainment areas (Gorman and Solomon 2002). Starting with the idea of "offset" credits in 1976, the EPA created local markets in which a new source could increase its emissions by paying an existing source to lower its pollution an equivalent amount. These credits eventually led to similar policies allowing for the transfer of emissions between stationary sources within particular nonattainment areas (Hahn 1989). Early emissions trading programs were unpopular with environmentalists, who desired a more aggressive approach to reducing emissions from the dirtiest sources (Doniger 1985; Nichols 1996). They were also quite unlike the idealized policies described in early economic accounts of cap and trade by authors such as Dales or W. David Montgomery, and tended to work with only limited success (Nichols 1996; Pérez Henríquez 2013, chap. 4).

The new EPA policies also followed the Lockean allocation norm, giving existing sources a valuable property right to emit in a given airshed in ways that discriminated against new sources (Gorman and Solomon 2002; Hahn 1989). Once applied to the air pollution case, the Lockean norm continued to influence allocation with little notice or debate in a series of emissions trading programs developed in the 1980s. In the successful program to trade rights for adding lead to gasoline in the 1980s (Hahn 1989), for example, there was no consideration given to other approaches for allocating the tradable permits: "After all, refineries had never paid a fee before and there was no reason even to think in those terms. As with the right to emit air pollutants in nonattainment areas, past use served as a baseline by default" (Gorman and Solomon 2002, 307). Thus, as cap-and-trade arguments drifted from their Pigouvian roots asserting the polluter should pay into more general efficiency-based claims, policy makers stuck with the more familiar Lockean norm grandfathering emissions rights for free to current polluters—a norm already used to guide allocations of many other natural resources by the government, including land and water rights. Misinterpretations of the Coase theorem may have contributed to this pattern, too, by suggesting that the initial allocation did not matter because the rights would end up with the same actors who valued them most highly regardless of how the rights were initially distributed.

Many have explained the pattern of grandfathering emissions rights in terms of theories of interest group pluralism (e.g., Huber 2013; Pérez Henríquez 2013; Cook 2010; Heinmiller 2007; Ellerman et al. 2000, chap. 2; Keohane, Revesz, and Stavins 1998; Hahn 1989). Grandfathering favors large emitters with more at stake who will expend substantial political energy to ensure they retain their low-cost emissions sink, thereby creating a barrier to entry (in the form of costly emissions allowances) for new potential competitors (Svendsen 1999). In addition, giving away allowances enables political actors to smooth the path for a new pollution control policy by using allowances as "side payments" to powerful constituencies, again consistent with interest group theories (Joskow and Schmalensee 1998). Finally, free allowances make the costs of a new environmental regulation less visible to the public—something that is appealing to large energy generators (Stavins 1998).

The history of early emissions trading programs is largely consistent with this interest group perspective and in conflict with the recommendations of economists that emitters should pay for their emissions. Regulators first adopted emissions trading reluctantly when faced with pressure to allow new economic development in areas failing to meet the CAA standards. As such, these officials were not interested in added controversy over allocation. Environmentalists were also initially skeptical of cap and trade, and focused more on limiting its use than on details such as how allowances would be allocated. Meanwhile, large emitters were motivated to assure that any allocation of emissions rights protected their existing use of the resource. The long history of using the Lockean norm to allocate other natural resources to existing users helped as well to make grandfathering the default choice for these new distributions. For all these reasons, the notion of auctioning emissions rights was never considered seriously in early emissions trading programs.

By the end of the 1980s, in sum, we see the basic outlines of the old model of cap-and-trade policy (table 3.1), resulting in a default allocation of grandfathering emissions rights for free to current users. For one, regulators were reluctant to create new private rights in public resources due to a lack of policy precedents, making them spend little time on questions of allocation. At the same time, environmentalists remained uninterested in allocation during this period, focusing much more on general concerns about the integrity of an emissions cap or moral concerns about creating a "property right" to pollute. Large emitters, by contrast, exercised the most influence over these policies and sought to protect their economic interests. The arguments that were made for auctions relied on efficiency-based considerations and failed to directly challenge the application of the Lockean norm that awarded ownership based on prior use favoring those large emitters. For all these reasons, the recognition of "squatters' rights" in allocation was dominant in emissions trading policy design prior to 1990, with auctions being truly "unthinkable" in most political discussions at the time.

Seeds of Change, 1990–2003

The old model remained influential over emissions trading policy right until the emergence of a new model for cap and trade under RGGI in 2003. Some challenges to the old model began to emerge in the 1990s, however, including in the design of the 1990 federal ARP to limit emissions of SO_2

Table 3.1
The Old Cap-and-Trade Model

- Regulators and environmentalists reluctant to use emissions trading; pay little attention to allocation

- Dominance of efficiency-based arguments for auctions

- Greater control of allocation rules by large emitters with large economic interests at stake

- Default to Lockean norm for allocation through grandfathering; auctions politically unthinkable

---

under the CAA Amendments of 1990. These various "seeds of change" (table 3.2) were important precursors for the more dramatic shift in cap-and-trade policy under RGGI.

During the 1990s, for example, regulators became more comfortable with cap and trade as a policy instrument, and began to experiment with alternative allocation rules. Like regulators, environmentalists in both Europe and the United States paid greater attention to allocation rules as they became more accepting of cap and trade as a pollution control policy. As a result, although new rules continued to give allowances away for free to emitters, they moved gradually away from the Lockean standard of grandfathering based on prior emissions levels. New arguments framing the allocation of emissions rights in terms of the polluter pays norm and (to a lesser degree) egalitarian norms of equal public ownership of the atmosphere also began to appear. Finally, other government policies started to experiment with auctions for previously unused public resources, such as new bandwidths on the broadcast spectrum, creating an important precedent for thinking about allocating pollution rights.

Electricity market deregulation also increased pressure for change to the old model in the United States. Deregulation threatened state rules forcing utilities to subsidize energy efficiency and ratepayer assistance programs, motivating environmentalists to seek new sources of funding for those programs. It also meant that states could no longer prevent power generators

Table 3.2
Seeds of Change Challenging the Old Model of Emissions Trading, 1990–2003

United States and Europe

1. Growing comfort and experience with cap and trade as a policy option leads policy makers and environmentalists to pay more attention to program design details

2. Initial foregrounding of application of Lockean norm to pollution rights, and first appearance of new polluter pays and egalitarian normative frames

3. New precedents of auctioning broadcast spectrum rights and severance taxes

Additional factors from electricity deregulation in the United States

4. Interest in finding revenue to replace public benefit funding for energy efficiency and ratepayer assistance programs threatened by deregulation

5. Stronger economic arguments against free allocation under deregulation

6. Fissures in political alliance among power generators and industrial energy consumers due to new competitive pressures from deregulation

from charging consumers for the value of emissions allowances even if those allowances were given to generators for free. This strengthened the economic case against grandfathering by increasing the risk of creating revenue "windfalls" for utilities. Lastly, deregulation helped create small fissures in the formerly united support for grandfathering among business interests due to recognition of the different economic implications of different allocation rules for different generators or industrial energy consumers, depending on their specific energy production and consumption patterns.

The remainder of this chapter documents the emergence of these seeds of change during the 1990–2003 period as well as their apparent influence on two important emissions trading programs developed just prior to RGGI: the US NO_x budget program, and the first phases of the EU ETS. These new developments led to a diversification of allocation rules beyond grandfathering, and even a onetime auction of NO_x allowances for new sources in Virginia. At the same time, these factors were insufficient for a policy change forcing *current* resource users to pay for most or all of their emissions, contrary to the Lockean allocation norm. The failure of efforts to successfully promote auctions in the EU ETS and NO_x budget program suggests that something was still missing that would make this dramatic policy change possible in the RGGI case.

Foregrounding the Lockean Norm in the 1990 ARP

The most famous example of a cap-and-trade policy for air pollution remains the 1990 ARP. Ending more than a decade of legislative stalemate over the problem of acid rain created by SO_2 pollution, these amendments to the CAA capped SO_2 emissions from large electricity-generating units at roughly half their 1980 levels, to approximately 8.95 million tons per year (McLean 1997). The ARP required affected sources to surrender one allowance for every ton of SO_2 emitted annually. Penalties for failure to surrender sufficient allowances included a fine and loss of allowances for future years. Subsequent assessments have indicated that the policy achieved a remarkably high compliance rate, with the benefits substantially outweighing the costs, making it a model for later emissions trading programs for GHGs (Schmalensee and Stavins 2012; US EPA 2009).

Consistent with the old model of cap and trade, the ARP faced resistance to creating a new "property right to pollute" (Burtraw and Palmer 2003; Cole 2002, chap. 3). In response, cap-and-trade supporters concentrated

on the environmental advantages of a cap on total emissions and the efficiency gains of allowance trading, expending little political energy on the
question of allocation (e.g., Dudek and Palmisano 1988). The regulated
nature of the electricity market at the time also limited concern about utilities gaining windfall profits through free allowance allocations—a crucial
difference from later allocation processes in the northeastern United States.
As a result, political actors distributed allowances for free to existing emitters, with some adjustments for new sources or sources in states with growing populations.

A remarkable aspect of this process is the continued lack of resistance
to the idea of giving rights away to existing polluters. Even new power
producers who received no allowances under the law did not object, asking
only for a chance to buy allowances from existing units. Although the ARP
included provisions for a modest auction, the sale affected less than 2 percent of all allowances and returned revenues to the firms that contributed
allowances to the auction pool. The auction was not a serious challenge
to the old model, in other words; it was simply a mechanism to improve
market liquidity (Stavins 1998; Ellerman et al. 2000). In sum, the idea of
charging firms to use this public resource appears to have been virtually
"unthinkable" at this time (Raymond 2003, chap. 3).

This is not to suggest that there was no interest in allocation in 1990;
most observers conclude the allowance allocation process determined the
bill's fate (e.g., Kete 1992; Cohen 1995). The political importance of the
allocation process is consistent with interest group politics arguments
claiming that free allocation is a useful device for obtaining the support of
key interest groups, including existing resource users, to ensure passage of
a new cap-and-trade policy. Along these lines, most published accounts of
the enactment of the ARP are grounded in the interest group perspective
(e.g., Cook 2010; Ellerman et al. 2000; McLean 1997; Stavins 1998; Burtraw
and Swift 1996).

A closer look, however, suggests that it is an oversimplification to assert
that the allocation was entirely driven by interest group influence. For
example, Paul Joskow and Richard Schmalensee (1998) test an interest-
based explanation of the 1990 allocation process in a variety of ways. Their
primary method is to calculate how different states would have fared under
different possible allocation rules, compared to the final allocation. The
authors are unable to find any obvious interest-based explanations for the

final allocation patterns, leading them to conclude that the allocation process was "more idiosyncratic than one might expect from previous work on the political economy of clean air" (ibid., 81).

Indeed, the results of the ARP allocation are contrary to the predictions of interest group politics theories: states with large numbers of relatively higher-emitting utilities did much worse under the final allocation than they would have under the traditional, pure grandfathering approach (table 3.3). The losses for these states are not trivial, and it is hard to see how one could argue that the costs were sufficiently "diffuse" in this case to weaken resistance to the final outcome by the affected utilities in these states and their representatives in Congress. Although other analyses (e.g., Bryner 1993, 144–146) have contended that a few special allocations to facilities in these states were important to making the final deal possible, table 3.3 suggests how relatively small those side payments were for states with the highest-emitting facilities. In fact, most large emitting states faced a substantial reduction in permitted emissions compared to a Lockean approach based on a prorated reduction of their 1985 emissions.

A more promising explanation for the initial move away from grandfathering is rooted in the strategy of foregrounding problematic norms, consistent with the model of norm-driven policy change outlined in chapter 2. During the debate over the acid rain bill, some participants began to

Table 3.3
ARP Final Allocation versus Grandfathering, High-Emitting States, 2001–2009

State	Actual allocation	Grandfathered allocation	Difference	Percent difference
Missouri	287,111	543,860	(256,749)	-47.2%
Ohio	686,279	1,254,440	(568,161)	-45.3%
Tennessee	266,257	453,724	(187,467)	-41.3%
Indiana	516,632	846,459	(329,827)	-39.0%
Georgia	415,525	564,753	(149,228)	-26.4%
Illinois	439,179	591,141	(151,962)	-25.7%
Totals	2,610,983	4,254,377	(1,643,394)	-38.6%

Notes: Units represent tons of SO_2 emissions per year for all affected sources in a given state. Grandfathered allocation represents prorated share of annual cap for all generators based on their actual 1985 emissions.

highlight the inapplicability of the prior use norm to the idea of pollution, decrying the awarding of valuable allowances based on a "right of prior *abuse*" (Raymond 2003, 89). Table 3.3 indicates that the designers of the ARP recognized this argument to some degree. Instead, Congress relied on the alternative concept of *benchmarking*: allocating based on what a facility would have emitted at a fixed emissions rate, applied across all sources, rather than the facility's actual emissions. Under the law, most allocations were calculated by multiplying an emitter's historical energy input for producing electricity (mBTUs of coal or natural gas) by an *equal emissions rate* that was far lower than the actual rates of many high-emitting power plants in states such as those listed in table 3.3. This new allocation approach represented an important first step in the erosion of the old model of emissions trading.

The adoption of benchmarking is not well explained by an appeal to traditional ideas of rent seeking or interest group politics. It is instead an early sign of the weakening of the Lockean norm influencing the allocation of this resource. As one individual who was closely involved in the ARP stated at a public meeting on RGGI in 2004, "Fairness of the emission rate was a key issue" in the design of the acid rain allocation (Greenwald 2004). Although Joskow and Schmalensee (1998, 63) hint at the significance of "equity arguments" in the ARP allocation, they appear to underestimate how criticism of applying the Lockean norm in this case made "true" grandfathering to large emitters politically unviable. Thus, policy impacts from criticism of the Lockean norm can begin to be seen as early as 1990. At the same time, new norms promoting the sale of allowances were not yet being applied to the issue, limiting the impact of this criticism to a weakening of the grandfathering standard.

Auctioning Spectrum Rights

Another crucial precedent for RGGI occurred around the same period that the ARP was being implemented: the auction of new broadcast spectrum rights for the first time in the United States. For decades, the US government had allocated exclusive use of various bandwidths of the broadcast spectrum to radio and television stations at no cost. Congress initially allocated these broadcast frequencies based on the familiar principle of prior use, whereby the first companies to broadcast on a frequency were given preference for formal licenses from the government. As with the grandfathering of emissions rights, this followed the familiar pattern of recognizing prior resource users in US law.

In 1927, Congress changed to a second model of allocation based on the idea of "public trusteeship" (Hazlett 1998). Under the Radio Act of that year, Congress agreed to grant broadcast licenses to companies at no charge in exchange for a commitment by those companies to operate in the "public interest"—addressing important public issues in their programming and providing equal time for different political points of view under the "fairness doctrine" (Moss and Fein 2003; Conrad 1989). Owners of licenses would periodically have to renew them through a hearing process to ensure they were meeting those public interest requirements. Most economists and many FCC regulators, however, dismissed these hearings as "beauty contests" with no clear standards and little to no chance of nonrenewal of an existing license (Cramton 2002; Holt 2006).

As with pollution, economists argued unsuccessfully for decades in favor of auctioning spectrum licenses (e.g., Coase 1959). Some also attribute the endurance of the free allocation system to traditional interest group politics (e.g., Hazlett 1998). On this account, by protecting the interests of current licensees and the power of key political actors with authority over the licensing system, the public trustee model of allocation was remarkably robust in the face of academic calls for reform.

In 1981, Congress adopted a new approach to allocating licenses for previously unused broadcast spectrum rights. The reasons for this change remain disputed, but a common explanation is that Congress balked at the impracticality of asking the FCC to allocate more than fourteen hundred new spectrum licenses (as required for new mobile phone technology) through a public trusteeship hearings process. Still reluctant to actually sell the licenses, Congress instead adopted a lottery to allocate them among all qualified applicants for the new bandwidths (Kwerel 2006).

The first lottery applied only to the allocation of new, currently unused bandwidths, and not for radio or television broadcasts (Hazlett 1998). In this respect, the new solution did not make current users pay for a resource they formerly enjoyed for free, thereby violating the norm of private entitlement based on prior use. Although part of the policy conversation, the idea of auctioning even these new bandwidths was too threatening to existing broadcasters and those supporting the public trusteeship doctrine to be politically acceptable.

The implementation of lotteries in the 1980s provided evidence of how much revenue Congress was forgoing by not auctioning these licenses. Numerous firms were created for the sole purpose of entering the spectrum lotteries, and some winners immediately resold the licenses on the open market at a substantial profit (Holt 2006; Kwerel 2006). In addition, concerns about federal budget deficits combined with evidence of large financial windfalls to spectrum lottery winners in the 1980s increased pressure to consider auctioning at least some spectrum rights, culminating in Congress giving the Federal Communications Commission (FCC) the authority to sell new spectrum licenses for the first time in 1993 (Kwerel 2006).

The result was a series of auctions of new portions of the radio spectrum to phone companies in the 1990s and early 2000s that raised more than $14.5 billion and were widely perceived as a success (ibid.). Several European nations followed suit, auctioning a variety of cellular bandwidths, some for substantial sums of money including more than $34 billion for a single spectrum auction in the United Kingdom in 2000 (Binmore and Klemperer 2001). Congress expanded the auction program in 1997, but continued to exempt license renewals and broadcast licenses from the program despite political pressure to auction those licenses as well (Hazlett 1998).

Growing experience with spectrum license auctioning in the 1990s is another important precursor to the successful push for allowance auctions in the RGGI case. The program represented a new foray into selling rights to use public resources rather than giving them away. It raised substantial sums of money and was cited as an example of how auctions could work successfully in deliberations over RGGI. At the same time, spectrum auctions were limited to bandwidths *not in current use*, consistent with the Lockean norm and interest group theories asserting the difficulty of making current resource users pay for their access rights. Although spectrum auctions were a big step toward the "RGGI revolution," they did not take the far more challenging leap of making firms pay for their existing use of a common resource.

Electricity Deregulation and Public Benefit Charges

From 1995 to 1996, the Federal Energy Regulatory Commission promulgated and enacted Open Access Order 888, allowing states to restructure their electricity generation markets. In response, all the states that would later participate in RGGI chose to deregulate their electricity markets between 1996 and 1999, trying to create greater competition among suppliers of electricity in order to spur innovation and create lower prices for consumers (Farrell 2001b). This change would have far-reaching implications for the politics of emissions allocation.

Most important, deregulation meant regulators could no longer prevent power generators from charging consumers for the value of allowances. In regulated markets, states could limit the ability of generators to profit from these free allocations by controlling the retail price of electricity. In a deregulated market, however, firms were free to set their own rates, and would incorporate the value of allowances in their electricity rates, even if the firms received them for free. This change meant that grandfathering allowances now threatened to create large new profits for power generators and higher prices for consumers.

Electricity deregulation also endangered programs prized by environmentalists that required regulated utilities to set aside money to promote energy efficiency among their customers (ibid.). With deregulation, regulators could no longer require utilities to provide these services in a newly competitive electricity market (Kushler 1998). As a result, many states responded by creating new taxes on consumers' electricity bills generally referred to as "public benefit charges" (Glatt 2010; Farrell 2001b). Public benefit charges (also sometimes called system benefit charges) represent a small tax (typically less than one cent) on every kilowatt-hour (kWh) of electricity used to fund a variety of energy efficiency programs as well as low-income ratepayer assistance (Glatt 2010; Bollinger et al. 2001). In many instances, the charges were set at a level to maintain existing expenditures on these public benefit programs in the years prior to deregulation.

Conflict over public benefit charges in the 1990s anticipated major arguments in subsequent debates over auctioning emissions allowances. Environmentalists noted the cost-effectiveness of efforts to reduce electricity demand through efficiency improvements, and maintained that those programs should be funded by a dedicated revenue source linked to electricity use (e.g., Galizzi et al. 2006; NY Public Service Commission 1996, appendix D, 28–29). Power generators and large consumers of electricity disagreed, contending that the private sector or general state revenues were more appropriate sources of funding for ratepayer assistance programs (e.g., NY Public Service Commission 1996, appendix D, 13–14). State regulators justified the new public benefit charges in terms of the need to allocate the costs of these programs "fairly" among all power generators as well as their ability to lower electricity costs for ratepayers, much as auction advocates would justify selling allowances in the future (ibid., 61; MA Division of Energy Resources 2007).

Despite being relatively small, these charges represented a "mini carbon tax" to fund programs to reduce the environmental and social impacts of electricity generation—an important political development. Even though utilities had previously incorporated the costs of energy efficiency and low-income assistance programs in their rates, those costs did not appear as a separate per kWh charge on utility bills. The decision to add such a tax to the bills of nearly all electricity buyers under newly deregulated markets was a significant step toward the idea of making the generators pay for emissions allowances in order to fund such programs under RGGI.

Finally, deregulation created a new economic landscape for power generators, putting firms in greater competition. In a competitive electricity market, lower-emitting sources of electricity had new incentives to support tougher emissions rules that gave them an advantage over competitors with higher-emitting power plants (Farrell 2001b). In addition, deregulation began to divide the interests of large industrial electricity consumers from those of generators, as customers sought the cheapest power possible on the grid. These changes did not completely undermine the opposition of power generators and large industrial customers to tougher emissions regulations, or to the idea of forcing firms to pay for their emissions, but they did start to undermine political alliances opposed to the idea of new allocation rules favoring cleaner energy sources.

First Emergence of Egalitarian Norms: Severance Taxes and the "Sky Trust"

As state and federal governments began to experiment with these small energy taxes, a few advocates also started to highlight larger state severance taxes as a model for future pollution control policies. *Severance taxes* are government charges on the development of minerals or other natural resources within a state, and are common in many western states with large oil, natural gas, and mineral reserves (Povich 2015; Rabe and Hampton 2015). Alaska has the oldest and most famous of these severance taxes, created in 1982. It taxes all oil extracted in the state, and puts much of the money into a statewide trust fund that generates an equal cash payment to every Alaskan citizen annually. Payments from the fund have averaged approximately $1,000 per person for most years of the program since its creation in 1982 (AK Department of Revenue 2014). Other oil-rich states such as Wyoming and Texas also have long-standing severance taxes, but do not follow the Alaskan model of putting the bulk of the revenue into a trust fund for citizens.

Although not strictly an auction of public use rights, these severance taxes are another important model for auctioning emissions allowances. The policies assert greater public rights over public resources, requiring private firms to pay for their access to those assets. Moreover, programs like the one in Alaska dedicate much of that revenue directly to the public, consistent with the public benefit model used to support auctioning allowances. Indeed, the Alaskan program was prominently cited in the late 1990s as a model for "cap-and-dividend" programs to charge polluters for their GHG emissions, and then distribute those funds equally to all citizens (Barnes 2001). According to this work, the Alaskan trust was a perfect model for a similar "sky trust" that would tax users of the atmosphere, conceived of as another public resource like oil or gas reserves, and then return those revenues to all citizens on an equal per capita basis (ibid.).

The sky trust idea differed from a traditional severance tax in some significant ways, not least of which being that severance taxes mostly raise energy prices for *out-of-state* consumers, making them politically more appealing. At the same time, the Alaskan model captured both of the key norms in the emerging public benefit model: making private users pay for their use of a public resource, and distributing those benefits in an egalitarian manner to most or all citizens. The sky trust idea clearly influenced some advocates working on the cap-and-trade-with-auction proposals for RGGI (e.g., Barnes and Breslow 2003), and Peter Barnes's (2001) idea of cap and dividend has become part of the language describing the new model of cap-and-trade policies that emerged in the early 2000s.

Allowance Auctions in the NO$_x$ Budget Program

One emissions trading program that developed in the wake of the 1990 ARP tried to limit emissions in the eastern United States of NO$_x$—an important contributor to ground-level ozone (i.e., smog) as well as acid rain. Although the EPA regulated NO$_x$ under the original CAA in 1970, the focus was on local emissions rather than regional and longer-distance effects. By the

1980s, greater recognition of the problem of long-range ozone transport led to pressure for new policies to address the movement of NO_x compounds on a regional scale. In response, the 1990 CAA Amendments included new NO_x regulations such as the creation of the Ozone Transport Commission (OTC) composed of twelve northeastern and mid-Atlantic states with persistent problems meeting federal standards for ozone (Aulisi et al. 2005). Most OTC states agreed on a regional emissions trading program as a significant part of their plan to reduce NO_x, and eventually the program expanded to include additional states (under what was called the NO_x State Implementation Plan [SIP] Call program) in 1999. Taken together, the OTC and SIP Call programs are often referred to as the "NO_x budget" program.

As a multistate effort to find new solutions to a regional air pollution problem in the Northeast, these NO_x cap-and-trade programs were important predecessors to the similar regional proposal for controlling GHG emissions under RGGI. This section reviews the process of determining NO_x allocations, including the extremely limited use of auctions, as an instructive comparison to the successful challenge of the Lockean allocation model in the RGGI case in chapter 4.

The NO_x OTC Program, 1994–1999

The OTC states agreed in 1994 on a memorandum of understanding (MOU) committing to a 75 percent reduction in NO_x emissions, primarily from large generators of electricity (Farrell 2001a). Emissions reductions were to take place in three stages, relying first on traditional command-and-control rules while states worked out the details of an emissions trading program to meet more stringent reduction goals. The MOU divided participating states into three zones, with different expected levels of NO_x reductions, based on their relative contributions to the most severe areas of ozone pollution in the major metropolitan areas along the Eastern Seaboard from Washington, DC, to Boston (Farrell 2000). It allocated NO_x emissions among the participating states based on emissions benchmarks per unit of prior energy production, much like the allocation of allowances to affected sources under the ARP (Carlson 1996).

Although the MOU determined each state's permitted amount of NO_x emissions, it allowed states to decide how to meet their particular emissions goals, including whether to participate in a proposed regional emissions trading program. Nine of the twelve OTC states chose to participate in the NO$_x$ trading program, with three states staying out due to either their small number of emissions sources (Maine and Vermont), or having already achieved the required standards on their own (Virginia) (Aulisi et al. 2005). The MOU gave participating states full control over the allocation of allowances to specific emitters in recognition of state concerns about giving up authority on this issue (Farrell 2001a; Burtraw and Szambelan 2009).

After completion of the MOU, a task force organized by two regional pollution control agencies (the Northeast States for Coordinated Air Use Management and the Mid-Atlantic Regional Air Management Association) began working on a model rule for states participating in the cap-and-trade system. The final model rule published in May 1996 continued to leave specific allocation rules to the participating states. From 1996 to 1998, states enacted their own NO$_x$ program trading rules based on the model rule, including rules for allocation to affected pollution sources (Aulisi et al. 2005), in a process that is remarkably similar to the one conducted to generate the cap-and-trade system for GHG emissions under RGGI ten years later.

Observers have noted the "difficulty" of determining NO$_x$ allowance allocations to individual emitters, consistent with interest group politics predictions for this sort of cap-and-trade policy design (e.g., Farrell 2000). State allocation processes varied, with some using legislation, and others allocating through regulatory rules (Aulisi et al. 2005). In the end, participating states adopted an array of allocation rules reflecting more incremental revisions to the grandfathering approach. Some chose to "update" their allocations over time based on future levels of plant activity, for example, rather than fixing the allocation permanently as was done in the ARP (Pérez Henríquez 2013, 155–156). In addition, at least two states moved to an "output" form of allocation where the relevant benchmark was based on energy produced by the power plant as opposed to the fuel consumed (Ellerman 2004). This innovation rewarded more efficient power sources by breaking the link between fuel used to make electricity and allowances awarded. Finally, some states also engaged in a fair amount of ad hoc allocation based on negotiations between affected sources and the regulating bodies (Krolewski and Mingst 2000; Sliwinski 2013; Seidman 2014).

This continued diversification of allocation rules reflects how the growing comfort with emissions trading among regulators led to a corresponding increase in attention to allocation options. Individuals involved with the RGGI program cited this incremental step in the NO$_x$ program as the "beginning of the idea" (Svenson 2011) that there could be different allocation approaches, and an important step toward auctioning under RGGI (Burtraw 2013; Lamkin 2012; Seidman 2014). It also reflected the new divisions among electricity generators, who benefited in different ways from different allocation approaches in newly competitive electricity markets (Farrell 2001b).

Besides facilitating new allocation rules by dividing political opposition among generators, the process of deregulating electricity markets strengthened economic reasons against grandfathering. Some individuals involved in the NO_x OTC design process recall hearing or making arguments for auctions based on the greater potential windfall to generators from free allowances, especially in New York (DeWitt 2013; Sliwinski 2013). These arguments criticized giving away allowances as a subsidy for power generators (Harrison and Radov 2002), again foregrounding the inappropriate application of the Lockean norm to this allocation context. Instead, new portrayals emerged of these rights as valuable public assets (DeWitt 2013)—the first signs of the new normative framing that would play a central role in later arguments for auctioning under RGGI. This was another big step in the model of norm-driven policy change outlined in chapter 2: offering the first alternative normative frames to replace the weak existing frame.

Discussions in the NO_x OTC program did not include detailed consideration about how to use auction revenue, however. In this sense, advocates were still offering only a limited new normative frame—one focused more on the importance of making polluters pay than on generating broad public benefits. Several individuals involved in the NO_x program design confirmed that these new ideas had only minimal influence on allocation designs for the OTC programs (ibid.; Sliwinski 2013; Bradley 2013), perhaps due to the lack of the additional focus on the distribution of these funds. Indeed, there is no evidence that any state took auctioning seriously in this initial round of discussions about allocating NO_x allowances from 1996 to 1998, despite the seeds of change present.

The NO_x SIP Call Program, 1999–2004

Even as states began implementing the OTC program, the EPA was working to expand NO_x controls to additional states throughout the eastern half of the United States. This process started as an EPA effort to organize a larger group of states into the Ozone Transport Assessment Group (OTAG) for a "consultative process" to reduce emissions (Farrell 2001a). Despite some technical successes, OTAG could not agree on an acceptable level of NO_x reductions for the "upwind" states in the group whose emissions tended to concentrate in "downwind" states. Faced with the political failure of OTAG, the EPA turned to another option: requiring a smaller group of states to submit new plans that specified tougher emissions limits on NO_x, starting in 2003 (Aulisi et al. 2005). This phase became known as the NO_x SIP Call program because it required states to revise their state implementation plans for meeting air quality standards under the CAA.

Once again, the EPA encouraged a cap-and-trade approach, saying it would automatically approve any state plan that included emissions trading as an implementation mechanism for achieving the new NO_x targets (Farrell 2001a). Twenty-two states eventually chose to participate in a regional cap-and-trade program for the NO_x SIP Call, including all the original OTC states (Napolitano et al. 2007). Unlike the OTC process, the EPA provided a detailed model rule for the SIP Call trading program that included allocation rules for emitting units (ibid.). Much like the ARP, the model rule allocated allowances to affected units based primarily on emissions benchmarks per unit of fuel input (US Code of Federal Regulations 2013).

Although not required to follow the model rule's allocation provisions, most states used the same benchmarking standard with minor adjustments (Burtraw and Szambelan 2009; Napolitano et al. 2007). One important modification created by some states again related to updating baseline fuel consumption information in calculating future allowance allocations (Napolitano et al. 2007). As in the case of the OTC program, some states also benchmarked allowance allocations based on emissions per unit of power produced rather than fuel consumed in order to reward more efficient production of power at a facility (Hibbard et al. 2000). These new forms of benchmarking continued the trend away from allocation based strictly on historical emissions.

Some states also set aside allowances under the cap for distribution to new sources of energy, as had been done in earlier emissions trading programs (Napolitano et al. 2007). Approximately six states withheld a small percentage of allowances from existing sources to be given to generators adopting energy efficiency or renewable energy programs instead of merely to new sources (Aulisi et al. 2005; US EPA 2005). Dedicating allowances to sources that reduced their emissions by investing in renewables and energy efficiency programs was another step away from the old cap-and-trade model. One individual involved with the allocation process noted that the set-asides for energy efficiency programs in Massachusetts were clearly motivated by the growing recognition of allowances as "valuable public assets" and the desire to use allowances to support important public benefit programs (Seidman 2014). To the degree that these energy efficiency programs used allowances to create subsidies benefiting all energy consumers, they moved beyond simply making polluters pay and toward generating broader egalitarian benefits. As a result, these set-asides were recognized at the time as an essential new element of these emissions trading programs, generating substantial public comment and argument (e.g., MA DEP 1999b).

Despite more active discussion of auctioning in the NO_x SIP Call program, including the new idea of allowances as public assets, the notion of auctioning such allowances still failed to gain much political traction (Napolitano et al. 2007; Burtraw and Evans 2003; Sliwinski 2013; Bradley 2013; Lamkin 2012). Interestingly, although the auction idea was percolating into these policy design conversations at this time, environmentalists were not making it a priority; comments by environmental groups on the Massachusetts NO_x SIP Call program design, for instance, failed to mention auctions at all (MA DEP 1999b, 4–5).

In addition, the limited discussions about alternative allocation rules stressed the importance of fairness to *generators*—leveling the playing field among various emissions sources—rather than possible benefits for consumers or the public from any new allocation plan. Out of hundreds of pages of documentation for the Massachusetts NO_x trading program, for example, there is only one mention of the program's relatively limited impact on ratepayers (MA DEP 1999a, 21). In this respect, any movement away from grandfathering was primarily driven by arguments about treating cleaner sources of energy fairly as opposed to impacts on the public at large—a pattern that was repeated in arguments over allocation in development of the EU ETS around the same period.

Thus, a benchmarked allocation based on updating was seen as "the fairest way" to handle the allocation issue at the time (Sliwinski 2013), with state regulators defending an output-based allocation in particular as creating a more "level playing field" for generators and being as effective as auctioning at encouraging the "most efficient investments" in pollution control (MA DEP 1999a, 15–19). At least one state also worried about its legal authority to auction allowances without legislative approval as well as potential competitive disadvantages if other states failed to auction their allowances in a similar manner (Seidman 2014; Levenson 2014). In the end, only one state conducted a onetime auction of a limited number of allowances under the SIP Call program—indicating the continued influence of the old model of cap and trade even in the NO_x budget program, where most of the seeds of change described in this chapter were present.

The Virginia NO_x Auction

Starting in 2002, Virginia chose to auction NO_x allowances set aside for new entrants—the first example of an auction of emissions allowances in which the state kept the revenue for public purposes. The state's Department of Environmental Quality (DEQ) had already proposed regulations in 2001 that would give NO_x allowances away to current emitters, consistent with other states and the wishes of large energy generators (Major 2013), despite the protestations of Department of Planning and Budget (DPB) analyst William Shobe, who suggested auctioning these allowances in his required economic impact analysis of the regulations. Even in making this argument, however, Shobe acknowledged that the auction would only work "as long as the charge is not enough to cause existing firms to effectively oppose the formation of the market" (VA DPB 2001, 11).

Yet in January of 2002, new governor Mark Warner confronted a serious budget shortfall and solicited ideas from the entire state bureaucracy for making up revenue (Paylor 2013; Menkes 2013; Shobe 2013a). Shobe (2006) responded to the governor's request by proposing that the state consider auctioning all its NO_x allowances, which he estimated would generate as much as $100 million in additional revenue. The governor's office was surprisingly receptive to this idea, possibly due to the fact that the governor had been active in buying and selling spectrum rights for cellular communications (Murray 2013; Shobe 2013a). Soon after, the Air Pollution Control Board suspended implementation of the proposed regulations giving the NO_x allowances away in order to provide time to explore auctioning at least some allowances. During this time the state's executive branch engaged in a series of "heated meeting, briefings, and discussions" about whether to auction some or all of the state's full complement of NO_x allowances (Shobe 2013b).

Although his 2001 economic impact analysis used more traditional economic arguments for auctioning, Shobe (2013b) defended auctions in 2002 using the new normative frame that these pollution rights should be treated as "valuable assets owned by the public" and therefore distributed in the same manner as other valuable public assets: by charging for their use. An official briefing on the auction proposal for the Virginia secretary of finance, for instance, included the following passage criticizing the proposal to give away allowances:

> In addition, these rules make a free grant to private firms of an asset that is used and valued by every citizen of Virginia. Charging for the use of this resource gives appropriate incentives for conserving it and serves as compensation to the public for giving up the good. This policy of charging for the use of the air and water for disposal services is sometimes referred to as the "polluter pays principle." (VA DPB 2002, 3)

Here we see another example of foregrounding the weak normative support for grandfathering allowances and the appearance of the new normative frame regarding the atmosphere as a public resource. Other sources also confirmed auction advocates using this specific frame in their arguments (Paylor 2013), and one former staffer recalled the governor asking at the time: Why would you give away a public good like these allowances? (Murray 2013). Here is initial evidence of the new public ownership frame leading elected officials to support the auction idea—a pattern that would become even more important in the RGGI case.

Even as the DPB was promoting the plan to auction *all* of Virginia's NO_x allowances, the state legislature began contemplating the idea of selling the much smaller group of allowances set aside for new sources. Having read about the market value of SO_2 allowances under the federal ARP, Virginia Senate Finance Committee staff member Neal Menkes proposed the idea of selling some of the state's NO_x allowances to meet the governor's mandate for new revenue sources. Unlike Shobe, Menkes (2013) focused on the smaller pool of set-asides because they were the ones that "no one had an entitlement" to, and so the legislature would "not be taking anything away from anyone," consistent again with the normative difficulty of challenging entitlements based on current or prior use.

After discussing the idea with staff at the Virginia DEQ, Menkes inserted a provision into the state's 2002 budget bill authorizing the department to auction the state's NO_x set-asides. The proposal cleared the relevant subcommittees and committees without much controversy, apparently based on the perception that these allowances were a "public resource" that had value yet were being given away at a time of potentially severe budget cuts (Menkes 2003). In this manner, conflicts in early 2002 over auctioning NO_x allowances in Virginia addressed both the larger DPB proposal to sell all such allowances as well as the smaller budget bill proposal to auction only the pool of set-asides.

Despite a positive reception from higher-level officials in the state budget office and the secretary of finance, both auction proposals faced opposition from the Virginia Department of Commerce and Department of Natural Resources, largely based on worries about the potential negative impact on private businesses and state economic development (Shobe 2013a; Major 2013; Burnley 2013). Others who opposed the auction idea also recalled the discussion being a "philosophical" one about the "right posture for the state to take vis-à-vis cost to industry" (Paylor 2013).

In May, opponents of the auction raised a new objection: that further delaying the allocation rules beyond July 1, 2002, in order to create new regulations for a full auction would violate the EPA's SIP Call deadline, putting the state at risk of financial penalties (Murray 2001; Shobe 2002). A memo to Governor Warner summarizing the pros and cons of delaying the NO_x regulations to make an auction possible directly acknowledges the new framing of the issue: "[Forgoing the auction] also potentially subjects the administration to criticism that it has given away a valuable public resources [sic] in a difficult fiscal climate free of charge. The biggest beneficiary, by far, would be Virginia Power, potentially allowing this to be framed as a giveaway to Virginia Power" (Murray 2001). Ultimately the governor chose not to further delay the NO_x regulations, effectively killing the idea of auctioning all the state's NO_x allowances. Possible reasons for the governor's decision cited by those involved in the process include: concerns about missing the EPA deadline, wanting to avoid antagonizing important state cabinet members and economic interests like Virginia Power, worries about the regulations hurting economic development in the state, or simply wanting to proceed more incrementally with a new policy idea and in the face of a Republican-controlled state legislature. Regardless of the precise motivation, the governor's decision not to pursue auctioning all the state's NO_x allowances is consistent with the outcome predicted by interest group explanations of allocation.

The governor's decision did not, however, eliminate the possibility of auctioning the smaller pool of set-aside allowances under the budget bill proposal. The DPB included $8.8 million in revenue from auctioning these set-asides in its fall 2002 budget proposal, and the final 2003 budget bill authorized a spring 2004 auction of these allowances with even more projected revenue (Shobe 2006). Interestingly, Senate staffer Menkes (2013) does not recall this budget process being particularly controversial, although power generators remained opposed to the idea. The Air Pollution Control Board narrowly approved new regulations to auction these allowances in December 2003, despite some grumbling within the board that it should not be "doing budget work for the general assembly," as one attendee at the meeting put it (Major 2013). Consistent with the new normative framing, most of the board's deliberations over the proposal apparently revolved around its "fairness" and potential environmental impact rather than issues of economic efficiency (Shobe 2013b). Just six months later, the actual auction of 1,855 allowances for both 2004 and 2005 NO_x emissions took place on June 24, 2004, raising more than $10.4 million in net revenue for the state (Shobe 2006).

The Virginia auction was a little-recognized landmark in emissions trading policy: the first nonrevenue neutral auction of allowances for air pollutants. At the same time, it represented a relatively modest and temporary change—one driven by state budget and finance employees pursuing new government revenue, and enacted through a budget bill instead of a standalone environmental regulation. Indeed, there appears to have been little excitement about the auction results at the DEQ, even as staff in the DPB celebrated the outcome (Major 2013). The final law authorizing the auction forbade any future auctions of NO_x allowances, indicating the exceptional and limited nature of this adoption of a new approach to allocation under the NO_x program. The Virginia program also took place largely outside the main political and advocacy networks that drove the development of RGGI's auctions, which were oriented around northeastern regulators and environmental advocacy groups that had little interaction with the Virginia process. Virginia's auction decision, in short, represents an interesting early step toward auctioning allowances, but one that remained relatively disconnected from the simultaneous initiation of RGGI in 2003.

In many ways, it is quite surprising that the larger proposal to auction more than $100 million of allowances got so close to being enacted, particularly in a state not known for its liberal or environmental politics. The unlikely confluence of a state budget crunch, an economist in the state budget office who was knowledgeable about emissions trading, and a governor who was experienced with the idea of auctioning other public assets like spectrum rights helps explain how this proposal to auction all allowances made so much progress. The reframing of these pollution rights as public assets subject to the polluter pays norm also appears to have played an important role in moving the idea forward politically, including obtaining support from the state's elected officials in the legislature and governor's office.

At the same time, those promoting the NO_x auction in Virginia failed to connect the polluter pays idea to an additional egalitarian norm demanding that use of this public resource must benefit a wide swath of the public. Lacking this additional egalitarian argument as well as the attention of environmentalists, auction supporters were still unable to overcome traditional opposition to the idea of auctioning from sources that would bear those costs.

Summing Up the Limited Use of Auctions in the NO_x Budget Program

The NO_x budget program was a critical precedent for RGGI. It involved many of the same states and a remarkably similar multistate negotiation process for creating the program's rules. It also featured nearly all the crucial seeds of change described in table 3.2 that emerged in the 1990s to challenge the old model of emissions trading, including the new economic arguments against grandfathering due to electricity deregulation. As a result of these factors, the NO_x program was the first emissions trading policy to use allowances as sources of revenue for public expenditures both by dedicating allowance set-asides to be used to support energy efficiency and renewable energy programs in a handful of states, and auctioning a small number of allowances in one state for government revenue for the first time.

Nevertheless, it is noteworthy how limited the discussion about and use of auctions remained in the NO_x budget program. Even foregrounding the strained application of the private entitlement frame to the allocation of pollution rights and presenting new frames describing how polluters should pay for their use of a public resource failed to overturn the core policy of giving nearly all allowances away to existing emitters. This continued reluctance to auction allowances indicates the durability of the old model of cap and trade even as the RGGI design process was getting under way in 2003. It also suggests that as important as electricity deregulation and growing comfort with market-based approaches were to spurring new allocation designs, those factors were insufficient for producing a major shift away from giving allowances to polluters at no charge.

Participants in the NO_x budget design process offered a variety of explanations for why auctions were not taken more seriously at the time given the changing economic and regulatory conditions. Some remarked that policy designers and public officials were not yet "ready" to understand the implications of deregulated energy markets for allocation, nor was there a sufficient community of "like-minded thinkers" to bolster such arguments (DeWitt 2011). In one telling example, environmental economist Dallas Burtraw (2011) recalled a question-and-answer session in 2003 with an audience of utility executives where the suggestion of selling allowances was received with guffaws. Others pointed to worries about insufficient legal authority in some states to sell allowances (Seidman 2014). In addition, many who worked on the NO_x budget program noted the more limited mobilization of environmentalists around the issue in this period as another important reason for the failure of auction proposals.

A comparison with the RGGI experience, however, will suggest another factor as being pivotal in the failure to generate auctions: lack of the right normative frame. This section has described how those promoting NO_x allowance auctions used new public ownership frames citing the polluter should pay norm, but failed to demand that the policy provide broad, egalitarian benefits to the public. In this respect, proauction arguments in the NO_x budget program paid relatively little attention to where auction revenues might be spent (DeWitt 2013; Bradley 2013; Sliwinski 2013; Shobe 2013a), consistent with the failure to include egalitarian norms of distribution in the new public ownership framing. Chapter 4 will illustrate how the addition of an egalitarian norm to the new framing in favor of auctions and specific policy designs supplying broad public benefits for ratepayers would prove critical in RGGI.

In sum, at this point a few economists were still working in relative isolation, trying to explain a fairly complicated economic idea to an unreceptive audience of regulators and elected officials focused on a vexing interstate pollution issue with fierce regional divisions. It is not surprising, then, that the NO_x OTC states continued to give allowances away for free, "presumably because of political resistance to auctioning," in the words of one review of the program (Aulisi et al. 2005, 19). Although these seeds of change that developed in the 1990s were sufficient to push some states to use more creative allocation rules designed to reward less polluting firms, they remained insufficient to break the taboo against making existing emitters pay for their allowances. Only a more active effort by environmentalists to promote the new public benefit frame combined with a new policy design based on that frame would make auctions politically viable.

Allowance Auctions in the Initial Phases of the EU ETS

Another program that developed in the wake of the ARP was the EU ETS, focused on controlling CO_2 emissions associated with climate change. Beginning operation in 2005, the EU ETS was the first emissions trading program specifically for GHG emissions. Despite efforts to promote auctioning, the European Union permitted member states to auction only a small percentage of their allowances in the initial phases of the ETS. Although the political and economic factors shaping the allocation debate in the European Union were different than in the NO_x case, several of the important seeds of change listed in table 3.2 were present, including increased comfort with cap and trade as well as the appearance of new polluter pays normative frames. Once again, however, these factors were insufficient for creating a major shift away from free allocation to current emitters.

Interest group theories appear to explain the decision to give away nearly all allowances in the first years of the EU ETS quite well. Early advocates for the ETS were economists working at the European Commission (the EU executive body), concerned primarily with generating support for emissions trading overall rather than on details such as auctioning allowances. Also consistent with the old model of emissions trading, environmentalists paid relatively little attention to allocation issues until later in the EU ETS process, concentrating at first on their misgivings about emissions trading in general. Large emitters, meanwhile, were active in promoting the emissions trading approach (Meckling 2011, chap. 5) and defending their access to free allowances. As a result of these and other factors, industry appears to have effectively controlled the early debate over the adoption and design of the EU ETS, driving the result toward a rejection of nearly any auctioning of allowances.

Initial Design Consultations, 1998–2001

It is surprising that the European Union took the lead on GHG emissions trading given that it opposed the idea in international negotiations leading up to the Kyoto Protocol climate change agreement. Once the Kyoto agreement was signed, the European Union's desire to make rapid progress on reducing its emissions combined with leadership changes at the European Commission's Environmental Directorate led the union to reconsider its stance and create the world's first emissions trading program for GHGs in 2003.

After the signing of the Kyoto Protocol in 1997, Belgian economist Jos Delbeke became the new leader of the Environmental Directorate. Delbeke and his staff favored market-based policies at a time when existing EU plans for energy taxes were looking less politically viable, generating a search for alternative policy approaches (Skjaerseth and Wettestad 2008, 74–75). Encouraged by powerful industrial interests such as BP and Royal Dutch Shell, the new commission leaders recognized the advantages of emissions trading as an alternative that could be more appealing to industry than an energy tax, while still using price incentives to achieve the union's new Kyoto emissions goals (Meckling 2011, 111–112).

The challenge of persuading reluctant member states and environmental interest groups to support emissions trading led the commission to pay little heed initially to the allocation of allowances. Its first official communication on the issue in 1998 merely stressed the need for emissions trading to be "part" of the way that member nations met their reduction goals (European Commission 1998, 19–24). Controversy over even using emissions trading continued to dominate the EU policy discussion through 1999, meaning that allocation continued to receive little attention in this period.

In 1999, the commission expressed more urgency about choosing specific policy mechanisms to meet the Kyoto goals. A new communication again promoted emissions trading as an important option, although only one that could be adopted after a "broad consultation exercise" (European Commission 1999, 15–16). This document also said little about allocation, except for a passing reference to the advantage of "benchmarking" as a method of distributing allowances within a member state that was consistent with EU rules against favoring domestic industries (ibid., 15). In recommending benchmarking, the commission illustrated the continued influence of the 1990 US ARP.

The commission also requested studies from economic consultants in the United States and Europe on possible emissions trading designs. Both consultants recommended auctions over free allocation, based on standard economic efficiency arguments. Both noted the likely political opposition to selling allowances, too, with one referring to grandfathering as "politically far more attractive" (Foundation for International Environmental Law and Development 2000, 31; see also Center for Clean Air Policy 1999, 18). During the same period, many European environmental groups continued to criticize emissions trading broadly rather than focusing on specific allocation details. Meanwhile, other research suggests that free allowances were the price of gaining acceptance of the policy from large emitters (Meckling 2011, 121–122). All this is in accordance with the old model of cap and trade based on interest group dominance of policy making.

Consistent with these developments, the commission's more formal green paper on emissions trading issued in March 2000 did not endorse allowance auctions. It instead continued to present the economic argument for cap and trade as the lowest-cost way to achieve emissions reductions guaranteed by the fixed emissions cap, while arguing for a cautious "step-by-step" approach to creating a new cap-and-trade program. By emphasizing the cost advantages of emissions trading and advocating the initial regulation of only large, fixed sources of pollution, the commission again followed the old model of cap and trade and the 1990 ARP (European Commission 2000, 9–14).

At the same time, the green paper praised auctions as an allocation alternative that was consistent with the polluter pays principle, and that gave an "equal and fair" chance to all companies to acquire allowances (ibid., 7–10). It also described how auction revenue could be used to promote energy efficiency efforts and research on low-carbon energy, or reduce other taxes and keep "the overall revenue effect neutral" (ibid., 18). Yet the commission failed to make any argument about public *entitlement* to the resource, or using the program to *benefit the public in an egalitarian manner*—additional parts of the new public benefit normative frame that were critical to the ability of RGGI activists to build support for selling allowances. Rather, the commission portrayed auctions as "technically preferable," noting likely objections by companies for having to pay "up front" now for "what had not been paid for in the past" (ibid., 18–19). In the end, the green paper made no recommendation on allocation, instead seeking input from stakeholders on various options as part of the broader consultation process.

The release of the green paper marked the beginning of an intensive discussion between the commission and various stakeholders. The commission held ten meetings between July 2000 and July 2001 with a working group of industry and environmental representatives to talk about the details of a final emissions trading directive (Meckling 2011, 117). In these working groups, more arguments emerged regarding the economic advantages of emissions trading as a low-cost mechanism for reducing emissions, including a presentation by the US EPA regarding the successes of the SO_2 and NO_x trading programs (Skjaerseth and Wettestad 2008, 83). Industry actively objected at these meetings to the idea of auctioning allowances, contending that the approach would threaten their competitiveness (ibid.). Meanwhile, environmental groups and their allies in the European Parliament remained more focused on other issues, such as regulating emissions "upstream" where the carbon entered the economy as well as limiting the overall reliance on emissions trading compared to other approaches to limit emissions (e.g., European Commission 2001b). Member states also remained largely indifferent to the idea of auctioning at this stage, with only Denmark making an argument for auctions based on the polluter pays norm (Danish Energy Agency 2000, 5).

A few environmental groups did advocate for auctions in their written comments on the green paper (European Commission 2001b). One group even invoked the polluter pays norm in support of auctions (Climate Network Europe 2000, 3–4), as did the European Parliament (2000, 51). The focus was not on the need for a broad benefit for the public, however, but rather on fears that grandfathering was unfair to some *emitters* by "distorting" competition among companies and penalizing sources that had already reduced their emissions (ibid., 51). Only one nongovernmental organization comment discussed the use of auction revenue, briefly mentioning its potential to help lower labor taxes, fund new climate mitigation efforts, or be divided equally among all citizens (German NGO Forum for Environment and Development 2000, 9). Another environmental group effectively spoke for most environmental commenters at the time in concluding that "the allocation method is not the most crucial aspect of the system as far as environmental effectiveness is concerned, and it should not dominate the process" (Climate Network Europe 2000, 4).

In sum, few stakeholders argued for auctions in the consultation regarding the commission's green paper on emissions trading. Those who supported auctions primarily based their reasoning on the idea that polluters should pay for their emissions and that auctions would avoid unfair outcomes for polluters that had already acted to reduce their emissions. There was virtually no discussion of the possibility of auction revenue being used to help the general public or ratepayers in a broad and egalitarian manner— only a focus on making sure all emitters were treated fairly and consistently with the polluter pays principle.

Introduction and Adoption of Formal Proposal, 2001–2003

The European Commission's draft version of the emissions trading directive, released in May 2001, gave member states the freedom to auction allowances if they wished. This permission to auction allowances generated a significant backlash from industry, resulting in a second round of stakeholder consultations. In these meetings, industry objected to making the emissions trading scheme mandatory, and argued in favor of "transparency" in allocation as well as more "flexibility" in the initial period of the EU ETS from 2005 to 2007, due to Kyoto targets not becoming binding until 2008–2012. A second consultation with member states also generated agreement on the desirability of requiring free allocations for emitters in all EU states, apparently based on industry pressure as well as concern about trade advantages for nations giving away allowances for free (Skjaerseth and Wettestad 2008, 120–123).

After this second round of feedback, the commission released its formal proposal for a trading system in October 2001. Giving in to industry pressure on some issues, the commission made the scheme mandatory for all member states, but recommended free allocation for all sources at least during the initial pilot phase from 2005 to 2007 (European Commission 2001a, 3–5). The proposal left open the possibility of new allocation rules for the second phase of the program, based on a review of the initial phase to take place by June 2006 (ibid., 11). The commission's main justifications for requiring a free allocation in phase one were the "experimental" nature of the first phase and lack of binding Kyoto targets before 2008, and the importance of a consistent approach across member states to avoid unfair competition (ibid., 5–6).

Surprisingly, the commission invoked the polluter pays norm in its justification for a free allocation, noting that emitters wanting to increase their
emissions would have to pay for those additional allowances (ibid., 6). Here
is another example of the limited ability of the polluter pays norm alone
to challenge the Lockean norm's justification of free allocation to existing resource users. In this case, the European Union applied the polluter
pays norm only to *new or additional sources* of pollution instead of existing
resource users, consistent with US broadcast spectrum auctions as well as
the 1990 ARP and Virginia NO_x auction.

The formal emissions trading proposal first moved to the EU Council
of Ministers, where allocation was not a major point of contention. The
"vast majority" of the member-states in the council apparently now agreed
that allowances should be given away "so as to stimulate participation"
by emitters (Skjaerseth and Wettestad 2008, 107–108). Things were different in the European Parliament, however, where the proposal moved for
consideration in spring 2002. Members of Parliament criticized the idea of
distributing allowances for free, and the rapporteur for the directive, Jorge
Moreira da Silva, stated his preference for auctioning 30 percent of allowances in phase one (2005–2007) of the program and 100 percent in phase
two (2008–2012).

Justifications for this higher level of auctioning continued to include
standard economic criticisms of grandfathering: making entry to the market difficult for new companies, inhibiting clear price discovery of the value
of allowances, and distorting competition among firms located in different
EU member states (European Parliament 2002b, 52). In addition, arguments
about needing to adhere to the polluter pays principle appear in the parliamentary record as well as concerns about grandfathering being "unfair" to
companies that had already made pollution reductions (ibid.). Still, there
continued to be no discussion of fairness *to the public* in the record, only to
emitters that had already acted to reduce their emissions prior to the program's starting date. The focus on a need for broadly distributed public benefits from any use of this public resource remained almost entirely absent.
Despite the arguments for auctioning, the Parliament's Environment Committee failed to amend the proposal to include an allowance auction—a
result attributed by one study to lobbying pressure by industry (Skjaerseth
and Wettestad 2008, 128).

At the subsequent reading of the directive before the full Parliament, however, environmentalists narrowly succeeded in obtaining an amendment calling for 15 percent of the allowances to be auctioned in the first and second phases of the ETS. The primary arguments in favor of auctioning allowances appearing in the parliamentary record continued to emphasize economic efficiency and fair competition among industries rather than benefits to the public (European Parliament 2002a). In this respect, many speakers favored stronger benchmarking standards for free allocations that would protect firms that had made early emissions reductions, consistent with new benchmarking standards emerging in the United States in the 1990s (ibid.). Even a Green Party representative (Mr. de Roo of the Netherlands) spoke in favor of auctions based on the idea that polluters should pay with no mention of the need to dedicate auction revenue to broad public benefits (ibid.).

The European Commission opposed even this limited degree of auctioning in the initial phases of the ETS, based on fears of industry opposition (Skjaerseth and Wettestad 2008, 130). At the same time, a newly elected coalition government in Germany insisted on free allocation as a condition of its willingness to support a mandatory trading system (ibid.). In response to these various pressures, the commission formally accepted some of the Parliament's amendments, but rejected the new auction provisions in November 2002 (ibid., 130–131). The European Council of Ministers soon followed by adopting a "common position" on the amended directive, largely consistent with the commission's revised proposal except for allowing up to 10 percent auctioning in phase two only (European Council 2003).

The council's adoption of only limited provisions for voluntary auctioning put the issue back in the hands of the European Parliament. Faced with pressure for a timely agreement in order to increase the chances of meeting the European Union's 2008–2012 Kyoto goals, the Parliament's Environment Committee amended the proposal to allow member states to auction up to 5 percent of allowances in phase one as well as up to 10 percent in phase two (Skjaerseth and Wettestad 2008, 134). Parliament then approved the amended text in July, the commission quickly accepted the new round of negotiated amendments, and the proposal with its limited provisions for auctions became EU law when adopted by the council as amended on July 22, 2003.

So why did auctions fail to take hold in the EU ETS, at least at these early stages? As in the NO$_x$ budget case, the seeds of change present in the EU ETS design process were insufficient to fully overcome obstacles to change in the old model of cap and trade. First, regulators in the European Union lacked experience and comfort with emissions trading, and were forced to spend much of their initial effort building support for the approach as a whole as opposed to focusing on allocation details. Second, environmental groups did not initially mobilize around allocation issues in the EU ETS, spending most of their early efforts objecting to emissions trading and trying to limit its use. Meanwhile, large industrial interests and power generators were attentive to defending free allocations based on status quo emissions levels.

Those supporting auctions also tended to rely on traditional arguments grounded in greater economic efficiency rather than creating broad public benefit from use of a public resource. Although the polluter pays norm appeared later in the EU process, it remained centered on treating emissions sources fairly versus providing important benefits for the public at large from the use of this public resource. In this respect, arguments for auctions in the EU case failed to take the crucial final step of connecting a new normative frame to the idea of creating a broad public benefit from auction revenue—the same pattern found in the NO$_x$ budget case. Absent a more aggressive normative reframing strategy connecting the polluter pays norm to egalitarian norms demanding broad public benefits from the policy, the EU adopted an allocation strategy quite similar to those of previous policies such as the 1990 ARP or NO$_x$ budget program.

At the same time, the creation and ratification of the EU ETS shows evidence of several significant signs of change. As the debate over the policy proceeded, comfort with cap and trade as a policy option increased, leading to greater attention to the policy's design details, including allocation. Environmental groups also displayed greater interest in allocation questions later in the process. Finally, the policy featured regular appeals to an important alternative norm for allocation: the polluter should pay idea. Although these changes were insufficient to create a major new policy design, they did contribute to the enactment of the first policy authorizing the sale of even a small number of GHG allowances to current emitters (even though almost no states actually auctioned any allowances in phase one of the program from 2005 to 2007; see Ellerman et al. 2010).

The EU process is also an instructive example of the influence that public agency staff members could exercise over the design of increasingly complex emissions trading programs. Although member states and others influenced the shape of the EU ETS, there can be little doubt that the ideas and arguments of the environmental staff of the European Commission had the greatest effect on the final design of the policy (Skjaerseth and Wettestad 2008, 142; Meckling 2011, 114). Chapter 4 will describe a similar pattern in the RGGI case, where public agency staff played a crucial role in the eventual decision to auction allowances.

4 Normative Reframing and the RGGI Revolution

By 2003, northeastern states were actively discussing a cap-and-trade program for GHGs as a response to federal inaction in this area. Following an extensive, multiyear policy design process with numerous stakeholder meetings and working groups, the RGGI MOU was signed by seven states in 2005, and adopted by three more in 2007. Although there was talk of covering a wider range of gases and sources, RGGI's designers ultimately settled on a cap-and-trade program for CO_2 emissions only from the electricity production sector, at least for the first period of compliance. The final version of the policy stabilized CO_2 emissions from affected sources through 2014, subsequently lowering the cap by 2.5 percent annually through 2018 (RGGI 2015a). RGGI states reaffirmed their commitment to the program in 2012, cutting emissions by another 45 percent.

In this respect, RGGI represents a remarkable example of a coordinated, multistate policy effort that changed the landscape of American climate change policy (Rabe 2008a). But RGGI is especially noteworthy for taking the groundbreaking step of auctioning nearly all emissions allowances. The first auction took place in September 2008, and through September 2014 RGGI (2015c) states had raised more than $2.2 billion from the sale of allowances, with an average price of $2 to $6 per ton. As noted in the book's introduction, commentators and participants in RGGI have already identified the allowance auction as the program's most important innovation. At a time when other emissions trading programs struggled to auction even a tiny percentage of allowances, RGGI was able to auction nearly all of them. By being the first program to make emitters pay for nearly all of their pollution, RGGI was truly revolutionary in the area of environmental policy.

RGGI thus can be seen as the vanguard of a new model of cap-and-trade policy design, in which environmentalists and regulators who are more comfortable with emissions trading have used the new public benefit frame to weaken the influence of large emitters over the allocation process, making auctions the new default allocation option. Table 4.1 summarizes this new model in comparison to the old model of cap and trade described in chapter 3.

The dramatic emergence of auctions caught many observers by surprise, including those closely involved in the RGGI design process (Hamel 2010; Murrow 2010; see also Rabe 2008a, 198). Combined with the fact that similar techniques have made auctions the default allocation option in other large cap-and-trade programs after RGGI, it is important to understand how exactly this revolution in emissions trading policy design occurred. How did RGGI's policy designers formulate and institute a new paradigm of public ownership of the atmospheric commons so quickly, even as efforts to promote auctions of emissions allowances in the NO_x program and early stages of the EU ETS failed?

This chapter considers that question in some detail. After reviewing the limits of existing explanations of RGGI's decision to auction allowances that are grounded in the interest group politics model, the chapter offers a detailed analysis of the RGGI design process to demonstrate the central role of the new public benefit model in making auctions politically viable. Public and private accounts confirm the prominence of this new normative framing for auctions, and its influence over the political process of making this policy change.

Table 4.1
Old versus New Model of Cap and Trade

Old model	New model
• Regulators and environmentalists reluctant to use emissions trading; pay little attention to allocation	• Regulators and environmentalists more comfortable with emissions trading; pay more attention to allocation
• Dominance of efficiency-based arguments for auctions	• Emergence of new public benefit frame for auctions
• Greater control of allocation rules by large emitters with large economic interests at stake	• Less control of allocation rules by large emitters in face of new normative frame promoting auctions
• Default to Lockean norm for allocation through grandfathering	• Plans to auction many or all allowances, and dedicate revenue to public benefits

Limitations of Existing Explanations for the RGGI Revolution

Prior accounts have agreed on the surprising nature of RGGI's choice of auctioning allowances, and largely agree that overcoming the opposition of vested economic interests was the key (e.g., Huber 2013; Rabe 2008a, 2010; Cook 2010). Many of these accounts also rely on explanations grounded in interest group theories of politics to explain this choice.

Several studies, for example, have noted that RGGI's decision to auction allowances was partially driven by the more liberal politics of the region combined with the participating states' long history of cooperation on pollution control issues (Huber 2013; Rabe 2010, 2008a, 2008b). In addition, high-emitting energy firms had reduced political influence in the RGGI process because northeastern states use relatively less coal for electricity production. And as mentioned in chapter 3, deregulation of electricity markets in the Northeast in the 1990s began to divide the interests of energy producers reliant on high- versus low-carbon fuels as well as the interests of energy producers, "wires-only" utilities no longer in the business of generating power, and large energy consumers (Huber 2013). Finally, the fact that out-of-state corporations owned many of these power-generating firms also made things more politically difficult for them (Murrow 2010; DeWitt 2011), consistent with the interest group model.

There is no question that these factors partly explain the adoption of auctions in RGGI. In particular, the relatively liberal politics of the region and lesser reliance on coal-fired electricity generation made the RGGI states an easier location for a major new climate change policy. It is nonetheless crucial to note that RGGI was a bipartisan program led and supported by several Republican governors, and that power generators remained politically influential and opposed to auctioning allowances in many RGGI states. More important, efforts to promote auctions made almost no progress in the earlier process of creating a cap-and-trade program for NO_x allowances that took place in these same states. For this reason, additional factors are required to explain how these states went from giving away all their NO_x allowances in the late 1990s and early 2000s to forcing firms to buy nearly all allowances for CO_2 just a few years later under similar conditions.

Previous accounts of RGGI's adoption of auctions have cited new policy entrepreneurs, new policy frames, new uses for auction revenue, and failures in the EU ETS as other factors critical to overcoming the objections of large emitters (summarized in the left-hand column of table 4.2). For example, many of these studies point to elected and appointed state officials as key policy entrepreneurs who made auctions possible in RGGI. Some authors emphasize the role of state governors such as Massachusetts' Deval Patrick or New York's George Pataki (Huber 2013; Cook 2010). Others stress the efforts of mid-level state bureaucrats (Rabe 2010; Engel and Orbach 2008; see also Rabe 2004), who had worked together for years on RGGI and previous regional air quality programs.

In addition, some of this work anticipates the main argument of this volume by discussing the importance of new issue frames. Barry Rabe, for instance, has pointed to the significance of framing successful state climate policies in terms of economic development. In extreme cases, what Rabe (2004, 2008b) refers to as "stealth" climate policies omit any references to climate change in order to enhance their political viability (see also Engel and Orbach 2008). Rabe (2008a, 2010) concludes that issue framing in terms of regional economic development was vital in RGGI.

Previous authors have also claimed that RGGI's policy designers used auction revenue to divide industrial opponents by allocating funds to certain firms to limit their opposition to the policy change (Huber 2013; Cook

Table 4.2
Competing Models for Explaining the RGGI Revolution

	Interest group politics explanation	Normative reframing explanation
(1) Key policy entrepreneurs	State governors and agency officials	Environmental advocates and (later) agency officials
(2) Main issue framing	Policy promotes regional economic development	Policy supports public right to broadly distributed benefits from use of public resource
(3) Use of auction revenue	Weaken industry opposition by dedicating auction revenue to programs benefiting some industries and not others	Increase political support by dedicating auction revenue to programs benefiting all energy consumers
(4) Key precedent	Problems with grandfathering in phase one of the EU ETS	Success of public benefit charges on utility bills
(5) Timing	Auctions introduced late in process; element of surprise	Auctions introduced early in process; major issue in all design discussions

2010; Tietenberg 2010). Finally, several studies have attributed the adoption of auctions in RGGI to an important new precedent: the collapse of the allowance market in phase one of the EU ETS. According to this explanation, auctions had a minor role in RGGI discussions until problems with the largely free emissions allocation in the EU ETS came to light in 2006, suddenly making auctions more politically viable (Huber 2013; Rabe 2008a, 2010). This introduces a final explanatory factor of *timing*: according to some accounts (e.g., Rabe 2010), auctions entered the RGGI policy discussion relatively late, catching business interests off guard.

Existing work on RGGI thus attempts to explain the adoption of auctions in a manner consistent with traditional theories of interest group politics. According to these studies, RGGI's elected officials and top administrators relied on the unique politics of their region, including their long-standing working relationships on regional pollution issues, as well as capitalizing on the public failure of the EU allocation process to introduce auctions late in the policy process. They promoted their new policy in terms of regional economic development, and used auction revenue to further divide potential opponents in the energy sector who were already less unified in their opposition to the program due to the deregulation of electricity markets described in chapter 3.

This chapter confirms that many of these factors contributed to making auctions possible in RGGI. Yet these explanations overlook several critical elements of the RGGI story, summarized in the second column of table 4.2. For example, previous studies neglect the role of nongovernmental advocates. This chapter will show how the idea of auctions originated from a small band of environmental advocates who decided to make auctioning a priority in RGGI, and struggled at first to bring reluctant state agencies on board. Although environmental advocates were also organized into their own interest groups, it is essential to underscore that their influence in this case is not well explained by traditional interest group theory. Rather than protecting their own concentrated interests through a micropolitical process, environmental groups had to identify and successfully promote a diffuse public benefit to replace the concentrated gains for power generators from free allowances, contrary to the expectations of interest group theories. Only when these policy entrepreneurs had convinced the relevant agency staff did auctioning gain momentum, as those government officials in turn helped convince agency heads and elected officials using the new arguments for auctioning that had been worked out during the first years of the process from 2003 to 2005.

Second, although policy reframing was indeed critical to the adoption of auctions, the most relevant frames were about the fair distribution of the costs and benefits of using the atmosphere as a "sink" for GHG emissions rather than about general economic development. This chapter will demonstrate the importance of the choice by environmental advocates to reframe the auction issue in this manner. These advocates continued to recognize that in the Northeast's deregulated electricity markets, consumer energy prices would increase the same amount under RGGI whether allowances were given away to energy producers or sold to them. As examined in chapter 3, several economists identified this problem even in discussions of RGGI's immediate predecessor, the NO_x budget program, to little effect.

By contrast, environmental advocates reframed allowances in RGGI as a public resource, not a private one, to be used for public benefit. Unlike more general frames for RGGI as an engine of regional economic growth (which auction supporters also embraced), the new public benefit frame was specific to auctions and the public's right to any value from private use of the atmospheric commons. In this respect, advocates complemented the argument for auctions based on the polluter pays norm with an additional claim promoting the broad distribution of any public benefits from the polluters' use of this resource, as explained in chapter 2. This reframing was a powerful political weapon; combined with economic analyses showing that deregulated energy producers would charge higher energy prices even if they were given allowances for free, it left power producers struggling to defend the practice of grandfathering. These new terms of engagement in the political conflict were far more important in weakening the political influence of energy firms than the allocation of a small number of free allowances to a few generators in order to reduce their opposition to auctioning.

Unlike those promoting these new frames in previous cases, such as the debate over the Virginia NO_x auction or design of the initial phases of the EU ETS, environmental advocates better tailored their policy proposals to match the new frame by describing auctions as "public benefit allocations," and dedicating auction revenue to energy efficiency subsidies or other measures for lowering consumer energy bills. In fact, a basic irony of the RGGI story is that an allowance auction offered a unique opportunity to address

potential public concerns about higher energy prices by charging generators for their allowances and then funneling that money to consumers to
lower their energy costs. Thus, arguments pointing to the use of auction
revenue to divide industrial and generator opposition to the program miss
the greater significance of using auction revenue to protect most consumers, consistent with the public benefit model.

In this regard, table 4.2 describes how public benefit charges were a more
important precedent for the design of RGGI than the initial problems with
the EU ETS. The use of auction revenue to support energy efficiency and
ratepayer assistance programs built directly on existing (and popular) public benefit funds along with the programs they supported. Without that
precedent, it is less likely that state agency staff and elected officials would
have been comfortable dedicating auction revenue to an entirely new set
of programs to help ratepayers and promote energy efficiency. Similarly,
the contention that auctions did not take center stage in RGGI until the
publicized problems with the EU ETS is also misleading. A careful look at
the early years of the RGGI process shows that auctions were a major part
of the earliest policy discussions dating back to 2003, well before the EU
ETS began operation in 2005. While subsequent problems with the EU policy added momentum to the push for auctions, those developments came
relatively late in the process after the decision to auction a large portion of
allowances had been taken.

This chapter documents the importance of these factors in spurring
the RGGI revolution toward dedicating private use of the atmosphere to
public purposes. First, it considers activity in the northeastern states that
led directly to the RGGI process, including prior state efforts under the
New England Governors/Eastern Canadian Premiers (NEG/ECP) Climate
Change Action Plan created in 2001. Next, the chapter looks at early planning stages for RGGI, starting with New York governor Pataki's invitation
to his fellow governors to participate in a regional program to control CO_2
emissions, and ending with the agreement on an MOU with a basic commitment to auction at least some allowances among seven northeastern
states in December 2005. Lastly, the chapter reviews the process of creating
a model rule for participating states in 2006 as well as the decisions of all
RGGI states to auction nearly all their allowances when the program began
in 2008.

Origins of RGGI

Responding to continued federal inaction on regulating GHGs, the NEG/ECP conference adopted a resolution in 2000 recognizing the threat of global warming and commissioning an action plan on the issue. After a March 2001 workshop, the environment and energy committees of the NEG/ECP presented the Climate Change Action Plan on August 28, 2001. Although the plan did not mention cap and trade, it set modest emissions reduction goals for the region and offered nine "action items" toward those goals, including new efforts to record and verify GHG emissions (NEG/ECP 2001, 8–18).

The NEG/ECP report instigated more intensive action among northeastern states. Massachusetts was the first to create a limited cap-and-trade policy for GHG emissions from power generators in 2001, with New Hampshire establishing a similar program in 2002 (Rabe 2008a). In April 2003, a New York task force recommended a wide range of policy options for reducing the state's GHG emissions, including a regional GHG cap-and-trade program for the Northeast (Center for Clean Air Policy 2003, ES-4-ES-7). The task force framed its recommended policies in two ways: as a new form of economic development for the state, and as a source of public health gains from the reduced emissions of pollutants associated with GHGs that are linked to human health problems (ibid., ES-9). (Interestingly, the task force also talked about economic development in terms of allowing the state to gain an early foothold in the expanding business of carbon-trading markets.) In this respect, New York's initial focus on economic development opportunities from reducing GHG emissions is consistent with the interest group explanation.

The report nevertheless also emphasized the success of New York's existing energy efficiency programs, including their ability to minimize the cost impacts of any new GHG emissions cap. Anticipating arguments that would be made by RGGI's auction advocates, the report stated that New York's electricity consumers could reduce their energy costs by participating in efficiency programs, even if energy prices rose due to a cap on emissions (ibid., ES-9-ES-11, ES-17). These arguments cited the success of existing public benefit funds to encourage consumer adoption of energy efficiency technology and recommend expanding those programs under any future policy. Despite the enthusiasm for public benefit funds, the report made no recommendation on how to increase funding for such energy efficiency programs or about how to allocate emissions allowances (ibid., ES-21-ES-23).

Outside the NEG/ECP process, however, new frames for allocating pollution rights were becoming more prominent. Anticipating the idea of public ownership of the atmosphere, New York attorney general (and future governor) Eliot Spitzer raised the subject of the "economic value of clean air" in his 2001 testimony on clean air regulations before Congress. Spitzer maintained that then-current regulations provided the atmosphere as a source of "free waste disposal" for polluters (US Senate Committee on Environment and Public Works 2001, 123). "Any normal company," Spitzer continued, "particularly power companies in the deregulated and highly competitive market, looks for ways to reduce costs. Free waste disposal, if allowed, is one such method" (ibid.). This testimony directly anticipated arguments in the RGGI process about the air being a public resource and that government needs to charge firms for using the public's air as a pollution sink.

Initial Planning, 2003–2004

In response to the recommendations of his GHG task force, Governor Pataki invited northeastern and mid-Atlantic governors to cooperate on a regional strategy to cut CO_2 emissions from power plants. By July 2003, eight other states agreed to work with New York on developing a program that would become RGGI. Meetings of representatives of the nine states began in September 2003 and would continue for the next several years on creating the details of this regional strategy, including how to allocate the rights to emit CO_2 under the program.

RGGI's design was developed through a lengthy stakeholder process. The Staff Working Group (SWG), made up of representatives from environmental and energy agencies in each state, managed the RGGI design process. SWG members reported regularly to their agency heads—the energy and environment commissioners. The SWG also created a public stakeholder group that included representatives of more than two dozen organizations, including utility, industry, and environmental interests as well as consumer advocacy groups. Starting with their first meeting on April 2, 2004, the stakeholder group met at least nine times prior to the release of the initial MOU for the RGGI states in December 2005. Stakeholder meetings continued in 2006 and 2007 on a less frequent basis.

The first meeting of the SWG on September 11, 2003, confirmed a group "consensus" to start with a program covering CO_2 emissions from electricity generation only (Litz 2003). Some representatives attending the meeting described RGGI as a "first step" in a broader effort to control GHG emissions; others were less sanguine about the idea of future expansion (ibid.). The group also organized task-specific subgroups, including one charged with drafting a model rule for all participating states (ibid.). These subgroups would do much of the heavy lifting on vital issues such as allocation, reporting back to the full SWG and stakeholder group on a regular basis.

The SWG grappled with the issue of auctioning allowances from the outset. As one environmental advocate closely involved with the process put it, "The allowances issue was a hot-button issue right from the very beginning—one of the most challenging issues [the SWG] had to deal with" (Bryk 2011). At first, environmentalists feared that SWG members would continue to follow the widely acclaimed 1990 ARP and take it for granted that allowances would be given away for free (ibid.; Murrow 2010; Kaplan 2011; Cowart 2011). State officials were skeptical of the political feasibility of auctions (Murrow 2010; Bryk 2011; Kaplan 2011), with one leading SWG member later describing the group's initial reaction to auction proposals as "not [having] even a remote possibility" of being enacted (Litz 2011). A state agency head also noted that "working in the penumbra of the 1990 deal on the acid rain program" made it harder to understand the arguments in favor of auctioning allowances (Campbell 2012). As one participant said of the process: "It felt like we were in a mental tug-of-war with the acid rain program for two years" (Cowart 2011).

This resistance shifted, though, in the face of the new claims by environmental policy entrepreneurs, who organized quickly within a month of Pataki's letter and prioritized auctions as "one of the things we could really do differently in RGGI" (quote from Murrow 2010; see also Kaplan 2011; Breslow 2011). Environmental groups including the Natural Resources Defense Council (NRDC), Environment Northeast, state Public Interest Research Groups (PIRGs), Clean Water Action, and the Conservation Law Foundation met regularly throughout the RGGI process, and coordinated their actions carefully from 2003 through the state approvals of RGGI in 2008 and beyond. Near crucial deadlines, these coordination efforts included weekly conference calls of twenty to thirty people (Murrow 2010), and environmental groups also consciously divided their efforts with different groups concentrating on specific states (Bryk 2011; DeWitt 2011). Despite a few disagreements about specific provisions such as whether to dedicate auction revenue to consumer dividends or energy efficiency programs (Breslow 2011), several environmental leaders observed that the coalition remained remarkably unified and organized throughout the entire RGGI process (Bryk 2011; Breslow 2011; DeWitt 2011). Interestingly, the advocacy group that first promoted emissions trading—the Environmental Defense Fund (2006)—was more peripheral to the RGGI debate and never fully embraced the push for auctions (Breslow 2011).

In an initial victory for environmental advocates, the SWG model rule subgroup's December 2003 outline of "key policy decisions" for the program included the question of whether to use auctions versus other allocation methods (RGGI 2003). The 2003 outline also considered the problems created for new generators by any allocation favoring existing sources as well as the potential for allocating to a much wider range of recipients, including nuclear and renewable sources, than had been considered in previous programs. Subsequent SWG meetings in January and February 2004 included several presentations from outside experts on different allocation options (RGGI SWG 2004b, 2004c; Litz 2004a). Within the first few months of the creation of RGGI, auctions were on the table, and the old model of cap and trade was under serious attack.

Allocation remained a major topic of conversation at the first public stakeholder workshop in April 2004. At this meeting, the chair of the model rule subgroup described allocation as a "key issue" and asked the group whether allowances should be allocated to sources other than existing emitters—an idea that would have been unthinkable under the old cap-and-trade model (RGGI 2004; RGGI Stakeholder Meeting Summary 2004a). At least one audience member noted that a "full auction" of allowances would solve many of the allocation questions raised in the presentation, bringing up the idea of auctioning 100 percent of allowances at the first stakeholder meeting to discuss the RGGI policy design (ibid.). Other speakers spoke of the advantages of auctions, while observing that they had been politically difficult in the past (Kruger 2004).

The first formal public comment submitted by environmental advocates also asked the SWG to consider auctioning up to 100 percent of allowances. The April 6, 2004, letter was signed by most of the major environmental groups active in the RGGI process, indicating the high degree of coordination among this group of advocates (Environment Northeast et al. 2004). The environmentalists' letter stressed that RGGI's economic modeling should include scenarios comparing the free allocation of allowances to auctioning all or a portion of them. Moreover, the group urged the SWG to consider the socioeconomic impacts of different allocation options, including effects on household income and generator profits, again anticipating arguments made under the new normative reframing about windfalls for generators and the need to protect consumers. The letter also suggested looking at the economic effects of giving auction revenue directly back to consumers as a rebate or using the funds for other public purposes such as energy efficiency programs. Here is evidence of environmentalists connecting the new normative framing to new policy ideas using auction revenue to create a larger benefit *for energy consumers*—one of the strategies missing in previous failed auction proposals.

Additional public comments from environmental advocates argued for auctions in May and June 2004 (e.g., NY Environmental Coalition 2004; MA Climate Coalition 2004). One of these comments, sent by many of the same advocates who signed the April 6 letter, noted that giving allowances away was likely to create "large new profits for some producers" of energy, and that auctions and free allocations would have the same effect on energy prices (MA Climate Coalition 2004). The group then presented a more detailed statement of the new frame of public ownership and polluters' responsibility to pay for their emissions for the first time:

> **There is no right to pollute**: Historically emitters of carbon dioxide and other pollutants have been allowed to pollute our air for free. There is no justification for continuing to allow "incumbent" emitters to have a greater right to pollute than others. The atmosphere, and the earth's climate are common property. All emitters of pollution should pay for contributing to air pollution and global warming. (ibid.)

The letter went on to justify using auction revenue "to minimize impacts on the general public," consistent with the normative reframing explanation in table 4.2. As in their April 6 letter, environmental groups advocated either a direct cash rebate to citizens or reducing energy costs by supporting energy efficiency programs as well as potentially providing "transitional assistance" to workers who lose their jobs due to the higher prices of fossil fuels. Finally, the group utilized a more traditional economic development frame, claiming that "auctions are far better for the [regional] economy" because they keep the value of allowances from leaving the region as profits for out-of-state energy company shareholders.

These comments confirm the importance of normative reframing in the strategy of environmental advocates working on the RGGI program from the beginning. Auction supporters who had previously focused on economic arguments were discovering the political power of the idea that "the right to degrade the commons is owned by the public," as one later put it (DeWitt 2011). In this approach, they consciously echoed ideas of public ownership of the atmospheric commons such as Peter Barnes's sky trust idea (Breslow 2011; Kaplan 2011) as well as existing public benefit charge programs.

The third RGGI stakeholder meeting in June 2004 was substantially dedicated to issues of allocation, and arguments connecting auctions to specific public benefits gained more attention. Speakers included Resources for the Future (RFF) economist Dallas Burtraw, who spoke directly about the "lack of fairness" in allocations based on historic baselines. Although he was not a formal member of the stakeholder group, Burtraw played an important role in promoting auctions in RGGI as a nationally recognized expert on emissions trading. Burtraw (2011) was convinced of the need for auctioning in the new world of electricity deregulation, later remarking that this was "one of those strange issues where equity and efficiency pointed in the same direction."

In his June 2004 presentation, Burtraw restated the usual economic point that generators would charge consumers for the value of emission allowances even if the allowances were given away for free. In the case of RGGI, he explained that the total allowance value could easily be two to twenty times larger than the actual compliance costs, creating large profits for those firms (Burtraw and Palmer 2004). Unlike previous economic discussions of the issue, however, Burtraw dedicated more than half his presentation to discussions of fairness and distributional impacts. First, he drew a parallel between auctioning pollution rights and government auctions of other resources, such as timber and spectrum rights, building on precedents reviewed in chapter 3. Next, Burtraw introduced the idea that a "stronger argument for free distribution is compensation for compliance costs." Thinking of allowances as compensation raises new policy questions, as Burtraw noted: How much compensation is enough? Are other parties are entitled to compensation besides existing polluters? (Burtraw and Palmer 2004). The underlying paradigm shift is profound: if current users are no longer entitled to free allowances based on prior use, someone else must effectively "own" the resource in question. That someone, of course, was the public, as environmental policy entrepreneurs had been advocating in their oral and written comments to the SWG.

Stakeholders at the workshop took up a similar refrain, with one speaker asserting that there should be "no sense of entitlement [to allowances] toward incumbents," and others framing auctions as an allocation mechanism connected to public benefits (RGGI Stakeholder Meeting Summary 2004b, 5–7, 13–14). As one commenter stated, "If we can't do auctions, we need to figure out a way to proactively distribute allowances to public benefit recipients. … We need to figure out how to provide generators with allowances to stay whole without providing windfalls to generators" (ibid., 7).

What continues to be noteworthy about this discussion is the attention to protecting the public—those likely to face higher electricity prices—from financial burden. Rather than simply making polluters pay, this new frame connected public ownership of the resource to the idea of auction revenue providing specific public benefits, including ratepayer relief. Consistent with the new frame, comments by environmental and consumer advocates after the June stakeholder meeting reemphasized the importance of reserving "most" or all allowances specifically for public benefit programs (Maine Public Advocate 2004; Redefining Progress 2004). Indeed, by this time most advocates were describing auctions as public benefit allocations, further reinforcing the link to support for the public at large. Consistent with the importance of earlier public benefit charges, environmentalists were also focusing more at this time on using auction revenue to support existing and new energy efficiency programs rather than for cash rebates for consumers (Bryk 2011; Breslow 2011).

Public comments by power generators and industry representatives during most of 2004 paid little attention to auctions. Instead, they concentrated on more traditional objections to environmental regulations, including the economic costs of limiting GHG emissions in the region (Center for Energy and Economic Development 2003a, 2003b; NY Coalition 2004) and a need to look at alternatives to a fixed cap on emissions (Northeast Regional GHG Coalition 2004b). A July 2004 Charles River and Associates study commissioned by industry underscored the "high economic burden" of limiting GHG emissions and raising electricity rates, with little attention to the issue of allocation or the claims of environmental advocates that auction revenues could be used to reduce demand and ameliorate potential electricity price increases (Center for Energy and Economic Development 2004). At the same time, economic modeling cited by environmentalists indicated RGGI would have a minimal effect on economic growth, and that investing auction revenue into energy efficiency programs improved the program's economic effects as well as benefiting ratepayers (Breger 2010).

In this respect, it appears that industry failed to recognize the momentum gathering behind auctioning in the first half of 2004, or at least failed to respond quickly. One utility representative later described the process as having been "hijacked" by environmental groups in terms of a sudden spotlight on auctions in 2005 (Svenson 2011), and others representing industrial or utility interests portrayed auctions as entering the process relatively late (Bradley 2011) and already being a "done deal" by the time industry offered their input (Rio 2011).

The relatively slow response from industry may also have reflected the modest initial emissions reductions under RGGI. According to one leading representative of utility interests, RGGI's cap remained high enough to make the "financial risks of auctioning manageable" for many firms, although less so for power companies relying more on coal (Bradley 2011). The fragmentation of interests in the generation sector, then, may have been another contributor to the delayed industry response, as firms with more lower-carbon power sources were more open to the change as a way to get a competitive edge (Rio 2011). It may have reflected the long-standing influence of the ARP and the old model of cap and trade, too, as some firms asked, "Why can't we just do this the way we did in the ARP and the NO_x programs?" (Bradley 2011). Finally, industry expended some effort to pressure state governors to withdraw from the program as a whole due to concerns about its economic impact, including the auction requirements, on power generators and ultimately ratepayers (Ruddock 2012).

Given the prevalence of discussions about auctions in the record from the beginning, it is hard to explain how industry failed to see the auction train coming down the tracks from the outset, except for the long history of auctions being politically unfeasible and therefore not worth worrying about. Whatever the reason, it was not until October 2004 that generators and industrial energy consumers began to argue directly against the idea of an auction, and by then at least some groups appeared resigned to the idea that auctions were going to happen and focused more on how to recover at least some auction revenues for their own benefit (Rio 2011; Ruddock 2012).

A full-day October 2004 workshop on Allowance Apportionment and Allocation increased the momentum behind the auction idea. Although some speakers dissented, the official workshop summary concluded that "a number of presenters indicated support for a glide path of increasing allocations for public benefit over time" (RGGI Stakeholder Workshop 2004a) and the allowances as compensation idea continued to gain traction. Emissions trading policy veteran Judi Greenwald (2004), for example, described the free allocation of SO_2 allowances in the 1990 ARP as "compensation" by noting, "Existing sources had to reduce a lot—so [it was important to] give enough allowances to match emission limitations." Dale Bryk of the NRDC went further, arguing for "public benefit allowance allocations" in RGGI to "reduce overall program costs" and "protect consumers." While Bryk (2004) was willing to countenance a "substantial cushion of free allowances" to power generators in the early years to cope with transition costs, she recommended an increasing percentage of public benefit allowances over time.

Just prior to the October workshop, power generators addressed auctions for the first time in their written comments. One group cited unfair competition with electricity generators outside the RGGI cap as a major reason against auctions, while another pointed to long-term contracts and other market factors as evidence that free allowances for emitters were not a windfall, and that emitters would not be able to pass along allowance purchase costs to consumers (Northeast Regional GHG Coalition 2004a; AES 2004). These written comments also argued that auctions would force many high-cost emitters to shut down, reducing the fuel diversity for electricity production in the region and jeopardizing system reliability.

Presentations and comments by electricity generators at the October workshop continued to stress unfair competition from firms outside the RGGI cap and the inability of energy producers to pass along higher costs from buying allowances to consumers (e.g., Braine 2004; Younger 2004). One even anticipated future controversies over spending public revenue from auctions by noting that allowances are "not a state budget fix" or "public benefit pocketbook" (Cunningham 2004). Although generators were now fighting back, they were on the defensive: the new public benefits frame was already at the center of the allocation conversation, forcing power producers to argue publicly against auctions in a new manner that put them at a disadvantage.

The October workshop appears to have been a milestone in the road toward adopting auctions. In a meeting just after the workshop, the SWG had a detailed discussion of various allocation options, including suggestions to allocate to zero-emission sources (as advocated by some renewable energy sources), allocate to new emissions sources, or be "fuel neutral" in allocation by giving all emitters the same number of allowances per unit of energy produced—an approach favoring production from cleaner fuel sources (Litz 2004b). In addition, the SWG appears to have discussed auctioning a small percentage of allowances at this 2004 meeting, with estimates of possible prices ranging from $5 to $10 per ton (ibid.).

Continuing to pressure the SWG, several environmental groups led by Environment Northeast submitted a recommendation for a "draft model rule" for the program in November 2004 that proposed auctioning 50 percent of allowances in the first year of RGGI, with revenue to be used for energy efficiency programs. The proposal suggested calling this a "consumer benefit" rather than public benefit allocation, apparently to further emphasize the ability of auctions to keep RGGI's costs to *ratepayers* low. The environmentalists' proposal also argued that auctions with revenue dedicated to reducing electricity costs were the best way to "ensure the program has the smallest impact on consumers," making several improvements on existing public benefit charges. Although the document still mentions consumer rebates as an option, it states a preference for spending auction revenue on energy efficiency programs to reduce consumer demand for electricity (Environment Northeast and Pace Energy Project 2004). By investing auction revenue in to expanding the energy efficiency programs described in chapter 3, these advocates claimed RGGI could reduce demand sufficiently to avoid any increase in the price of electricity from the new cap on GHG emissions.

From their perspective, the environmentalists' proposal appeared to set the terms of the debate for the next phase of the process (e.g., Murrow 2010). Power generators at this point concentrated more on auctioning, reiterating several arguments against public benefit allocations at a November RGGI

Stakeholder Workshop (2004b) including issues with long-term contracts and competition from lower-price electricity sources outside the RGGI cap as well as threats to the reliability of electricity supply. Although industry was now condemning the idea of auctioning more directly, it remained on the defensive. By December 2004, for instance, at least one leading industry group was supporting an "output-based" allocation that would provide the same number of allowances per unit of energy produced regardless of the actual historic emissions or type of fuel used—an idea that had also been considered in the SWG meeting in October (Northeast Regional GHG Coalition 2004c). This is the strongest evidence yet of the success of the new normative frame in undermining the practice of grandfathering. As early as December 2004, discussions of pure grandfathering had effectively vanished, and the debate was instead about what new allocation rule would replace it.

Creating the MOU, 2005

By early 2005, the SWG began planning for a series of presentations to its agency heads regarding a "Preliminary Design Proposal" for RGGI, covering issues ranging from the recommended cap level to possible rules for allocating allowances to provisions for offsets (RGGI SWG 2004a, 2005a). The first of these presentations would occur at an April 2005 conference at the Pocantico Center in Tarrytown, New York, where the SWG would review its modeling work and stakeholder input, and deliver a first draft of its recommended design principles for discussion and then subsequent revision and reconsideration at a follow-up meeting that summer (RGGI SWG 2005a). These meetings would serve as the first crucial decision point for the RGGI process, generating an initial recommendation to elected officials regarding allocation and other controversial design questions that would form the basis of an MOU among the participating state governors.

Stakeholders' arguments over allocation intensified as the SWG prepared for this vital series of meetings. A group of environmentalists led by the state PIRGs endorsed the general approach of the draft model rule submitted by Environment Northeast, but urged that even more allowances be dedicated to consumer allocation (Environmental Advocates of New York et al. 2005). Meanwhile, Environment Northeast continued to partner with a larger group in pressuring SWG modelers to consider the possibility of reducing energy demand growth to zero through the increased funding of energy efficiency programs with auction revenue (Environment Northeast et al. 2005b). The Maine Public Advocate's office also reiterated the claim that giving allowances away would raise electricity prices for consumers without compensation, and that allowances should be used instead "in ways that actually mitigate the impacts of the RGGI program on consumers" (Joint Utility Consumer Advocates 2005).

At an April 2005 stakeholder meeting, RFF's Burtraw presented another analysis of allocation options, again identifying the historical precedents for auctioning and focusing on the idea of allowances as compensation (Burtraw, Palmer, and Kahn 2005). Some auction advocates, including one author of the Environment Northeast model rule proposal, argued at this meeting against giving away *any* allowances to current emitters: "Consumers, who pay the price increases, should receive the proceeds of allowance sales" (DeWitt 2005). Comments from the audience echoed the importance of helping consumers: "We don't advocate auctions specifically, as long as there is a direct allocation to consumers" (RGGI Stakeholder Meeting Summary 2005a). The SWG had also begun publicly signaling that the environmentalists' message was getting through, saying at a February 2005 stakeholder meeting that it could see some allowances being used to fund energy efficiency programs (Tierney 2005).

A follow-up letter by many leading environmental advocates forcefully restated the economic and fairness advantages of auctioning allowances, reemphasizing the link to promoting energy efficiency:

> The RGGI program should be designed with consumers in mind and to maximize economic benefits. The economic value of allowances should be recognized by policy makers and [allowances] should be allocated in a manner that recognizes that consumers primarily bear the burden of the program and that investments in energy efficiency will reduce or eliminate that impact. (Environment Northeast et al. 2005a)

Environmental groups highlighted the political rationale for their approach: that elected officials would want regulators to have "ensured that the [RGGI] rule will impose minimum cost on consumers and provide net economic benefits to the region" (ibid.). In this sense, framing in the RGGI case was primarily directed at public officials, making the argument that the public *would* perceive auctions more positively than an approach using grandfathering, based on the new normative frames being promoted by environmental advocates. This is in contrast to the alternative normative reframing strategy, discussed in chapter 2, of promoting the new frames directly to the public in the hopes that people will then demand policy change.

Finally, the environmentalists maintained that their proposed public benefit allocation was technically not an auction but rather "an allocation of a *public good* to a regulated entity for the benefit of consumers" (ibid.). All these arguments by environmental advocates stressed the specific policy implications of the new normative framing that the public deserves any revenue from the privatization of the right to pollute.

The continuing distinction made by environmentalists between an auction and a public benefit allocation illustrates the significance of reframing to their strategy. Although it was partly generated to head off potential legal objections about states' lack of authority to sell allowances (Bradley 2005), the language shift went beyond legal hairsplitting. The idea of a public benefit allocation reinforced the normative reframing of allowances as a public asset as opposed to a private entitlement, focusing attention on the main idea that allowances must benefit the public rather than the concept that polluters must pay for their use of the resource. The key was not selling allowances per se but instead rebutting the presumption that emitters should get allowances for free and ensuring that the value of allowances goes to the public (Burtraw 2011; DeWitt 2011). In this respect, one advocate mentioned wanting to create an online "auction tracker" to allow the public to see where the money was going after every auction (Murrow 2010). Here again is an example of how simply insisting polluters should pay was not enough; connecting the notion of public ownership to the idea of creating a broad public benefit from any auction revenue was even more important.

Power generators continued to struggle to respond to this new normative frame, especially the claim that they would charge consumers for the value of allowances in their electricity pricing even if they received the allowances for free (DeWitt 2011; Burtraw 2011). Many continued to change the subject by asserting that states lacked sufficient legal authority for an auction, auctions would damage the region's economic competitiveness, or auctioning would threaten the stability of electricity production in the Northeast (Bradley 2005; Younger 2005). They also maintained that auctions were politically unrealistic as a model for a national program (Bradley 2005).

Lastly, they appealed in vain to the success of the old model of cap and trade as embodied in the 1990 ARP, contending that the "historic model" of grandfathering allowances had been used in other programs with "proven success" (Younger 2005). Contrary to previous experience, these arguments had little effect on the movement toward adopting a new approach to allocation in RGGI.

In April 2005, RGGI state agency heads met in Tarrytown to consider the initial recommendations of the SWG. The discussions appear to have revolved around the overall cap level, timing for implementing the program, and in particular the apportionment of allowances among participating states as well as the possible role of offsets (RGGI Stakeholder Meeting Summary 2005b, 5). The public benefit allocation idea was also introduced at this meeting, although the agency heads gave that issue greater attention later in the MOU development process (ibid.; Sheehan 2012; Campbell 2012).

As expected, the agency heads ended the Tarrytown retreat by asking the SWG to submit an updated proposal in the summer. Key aspects of the updated RGGI SWG (2005d) proposal included:

1) Starting the program in 2009, and capping emissions at current levels through 2015, and 10 percent below current levels by 2025
2) Permitting offsets from projects both inside and outside the RGGI region, but limiting them to covering no more than 50 percent of a given source's emissions
3) Allocating to states based generally on historic emissions levels
4) Letting states decide how to allocate to individual sources, as long as they "consider an agreement to allocate at least 20 percent of the allowances to a public benefit purpose or purposes," defined primarily as energy efficiency investments
5) The promotion of "complementary energy efficiency policies" in all participating RGGI states to further reduce energy demand and minimize any possible rate increase from the cap

Although it recommended a smaller percentage of allowances for public benefit allocations, the SWG proposal relied on many of the same justifications for the idea presented by environmentalists. The SWG described how greater investments in energy efficiency could reduce energy demand, thereby lowering energy prices, and followed environmentalists by referring to existing public benefit charges as an important precedent for the public benefit allocation (RGGI SWG 2005b). Although there is no direct mention of the idea of public ownership of allowances, the proposal's implication that the program should use auction revenue to benefit *all* ratepayers is consistent with the normative frame promoting that new idea.

In August, the SWG presented a final revision of its proposal to the agency heads. Although the heads did not finalize the recommended plan until later in September, the August draft was announced publicly and served as the basis of the MOU to be signed by the RGGI governors by the end of 2005 (RGGI Stakeholder Meeting Summary 2005c, 1–2). The revised proposal (RGGI SWG 2005c) had a few key changes from the July version: the 10 percent emissions reduction goal was moved up to 2020 from 2025; state allowance budgets were finalized based primarily on historical emissions with some adjustments on the basis of "electricity consumption, population, potential emissions leakage, and provision for new sources"; and the SWG added a new "strategic carbon fund" reserving an additional 5 percent of allowances to fund emissions reduction projects that would compensate for an expected increase in energy production and emissions outside the region due to higher energy imports into some RGGI states. Combined with the 20 percent of allowances dedicated to public benefits, this meant the final SWG proposal included recommendations for effectively auctioning at least 25 percent of all allowances—a startling accomplishment even in the eyes of environmental advocates who had fought for auctions from the beginning of the process (Murrow 2010; DeWitt 2011). Indeed, even after making its final recommendations in August, the SWG still recognized the public benefit allocations as "one of the more controversial aspects" of the RGGI model rule (RGGI Stakeholder Meeting Summary 2005c, 4).

Stakeholders on all sides mobilized in response to the release of the final draft plan. A critical development at this point was the decision of National Grid (2005a), a large electricity utility that did not generate power itself, to support the idea of auctioning allowances and dedicating the revenue to lowering ratepayers' costs. This public support for auctions by a utility was a watershed moment (Kaplan 2011; Murrow 2010; Bradley 2011). Environmentalists had worked hard to persuade the company of the problems with a free allocation, eventually leading key individuals at National Grid to question the assumption of giving allowances away (Kwasnik 2011). In response, the firm commissioned its own economic study confirming that grandfathering allowances would benefit generators in a manner that might be consistent with other political goals yet not help consumers (National Grid 2005b). The company thus chose to support auctions because it was "much fairer" for customers and "politically more palatable," as one former executive at the firm observed (Kwasnik 2011).

Environmentalists also kept up the pressure, repeatedly defending auctions on the grounds of consumer protection (Environment Northeast et al. 2005c; Joint Environmental Organizations 2005) and the "fairness" of requiring polluters to pay for their emissions rather than giving them away for free (National Association of State PIRGs 2005a, 2005b). Disagreement nevertheless continued over what fairness required in terms of using auction revenue. Most environmentalists by this time advocated using a substantial portion of auction revenues for energy efficiency programs versus cash rebates to ratepayers. National Grid and other business groups favored a cash rebate as the "fairer" option (RGGI Stakeholder Meeting Summary 2005c, 28, 38). This disagreement about the ultimate disposition of public revenues would gain salience in future auctioning debates, but was not yet a major point of disagreement in RGGI.

Forced to grapple directly with the new normative frame, generators began to challenge the SWG proposal in terms of its unfairness to high-emitting sources such as coal-fired power plants as opposed to only focusing on the plan's potentially negative economic impacts (Edison Electric Institute 2015, 14–15). This effort to argue against auctions in terms of *fairness to high emitters* was a difficult sell given how the new normative framing had foregrounded the weak fit between the norm of beneficial prior use and allocation based on previous levels of harmful pollution. As a result, this attempt to defend the old model of allocation in terms of fairness appeared to have had little impact and was soon abandoned.

Political leaders paid greater attention to RGGI and the allocation issue in 2005 as the MOU moved closer to completion, with the agency heads finally agreeing on most of the details of the SWG proposal at another meeting at the end of September (Sheehan 2012). Leaders of the SWG undertook a "mini-tour" of the region to meet with governors and their staff members that year to brief them on the basics of RGGI, and also raise the auction issue (Litz 2013; Hamel 2013). These political conversations shifted to a higher gear in the fall as the MOU was completed. At the September RGGI Stakeholder Meeting (2005c, 41), for example, a representative of the New

York Attorney General's Office urged the SWG to "consider a larger allocation to the public than 20 percent."

Some elected officials were quick to see the appeal of auctioning allowances, especially in terms of giving states money to do things that would be politically popular, including promoting energy efficiency (Kaplan 2011; Litz 2011). The presence and popularity of existing programs dedicated to improving energy efficiency in many RGGI states also made the public benefit allocation idea more viable (Campbell 2012; Sheehan 2012; Lamkin 2012). As one interviewee put it, the public benefit allocation "really resonated with the structure in place" in terms of the existing public benefit charge programs (Campbell 2012).

By the time the SWG proposal was finalized, however, at least one governor was having second thoughts. Under increasing pressure from industry, Massachusetts governor Mitt Romney began to waver in his support for RGGI, calling a meeting with business and environmental leaders in his state at the end of September (Novak 2006). The key issue for the Romney administration in the context of potentially withdrawing from RGGI in 2005 appears to have been a desire to cap potential allowance prices, however, rather than the auction of allowances (Ruddock 2012; Hamel 2013). Eventually, these concerns led Governor Romney to withdraw from RGGI in December, with Rhode Island following suit (Massachusetts Launches CO_2 Control Plan 2005; Avril 2005). Connecticut's Republican governor nearly withdrew as well at this time based in part on arguments from the Romney administration, but in the end Connecticut signed the MOU to the relief of the SWG and auction supporters (Litz 2011).

Thus, on December 20, 2005, seven states signed the MOU: Connecticut, Delaware, Maine, Massachusetts, New Hampshire, New Jersey, and New York. In the final document, these states agreed to dedicate 25 percent of the program's allowances for "consumer benefit or strategic energy purposes" (RGGI 2005). These various "purposes" were defined in the MOU as promoting energy efficiency, mitigating ratepayer impacts, promoting renewable energy technologies, or funding administration of the program itself (ibid.). Although the MOU never mentions the word auction, media reports used that term to describe the 25 percent consumer benefit/strategic energy allocations (e.g., "Seven Northeast States" 2005; Miller 2005). The MOU set the starting date for the program at January 1, 2009, specified the regional cap, and finalized allocations of the cap to participating states based roughly on status quo emissions (RGGI 2005). The MOU also included a "safety valve" that would extend compliance deadlines if the twelve-month average price of allowances exceeded $10 per ton, and sharply reduced the allowable use of offsets to only 3.3 percent of emissions (ibid.). In another effort to control compliance costs, a similar allowance price threshold moderately expanded the use of offsets to meet emissions goals under the MOU.

Although the loss of Massachusetts and Rhode Island was a blow, the signing of the MOU by seven of the RGGI states was a turning point in the pursuit of auctions under the program. As several interviewees observed, once political leaders came around to the new arguments in favor of auctions, momentum built to auction even more than the 25 percent required by the MOU (Murrow 2013; Litz 2011). One advocate put it this way: "The MOU language really broke the dam" on the issue (Kaplan 2011). Although the MOU left many issues unsettled, including exactly how the revenue from any public benefit allocation would be spent, the old model of grandfathering allowances to existing emitters was effectively finished once these seven states agreed in principle to sell a substantial portion of their allowances for public benefit purposes.

Developing the Model Rule, 2006

After the signing of the MOU, the SWG turned its attention to finishing a model rule for states to follow in drafting legislation and regulations for implementing RGGI. In this regard, if the MOU was like a skeleton of general principles for RGGI, the model rule was the first effort to put some "meat on the bones" in terms of exactly which sources would be affected by RGGI, how states would distribute allowances, and how they would deal with enforcement and offset provisions.

The SWG released a draft model rule for public comment on March 23, 2006. Largely consistent with the core elements of the MOU, the draft rule included a few new provisions exempting certain types of power generators from the program and offering more detailed standards for projects to qualify for offset credits (RGGI 2006). Although these provisions created new controversy, a primary point of contention remained the degree of reliance on auctions. The model rule was less specific than the MOU on the issue, saying only: "Allocation provisions will vary from state to state, provided at least 25 percent of the allocations will go to a consumer benefit or strategic energy purpose" (ibid., sec. XX-5.3[a]). Unlike the MOU, the draft model rule did not specify what qualified as a consumer benefit or strategic energy purpose, inspiring many requests for clarity about what expenditures would meet this standard.

Having succeeded in establishing the legitimacy of auctions in the MOU, auction advocates saw an opening during model rule discussions to press harder for their goal of 100 percent auctioning (e.g., Murrow 2010; DeWitt 2011). As noted above, they sensed that the hardest challenge of gaining acceptance for the new frame supporting auctions had already been accomplished in the MOU fight. "If you bought the rationale for auctioning any," said Franz Litz (2011), an SWG cochair, "then you really had to auction them all." In this sense, the new normative framing made anything less than 100 percent auction of allowances hard to justify.

The public comments by environmental groups on the draft model rule reflect this growing momentum (e.g., Nature Conservancy 2006; MA Climate Action Network 2006; ACEEE and Alliance to Save Energy 2006; NRDC 2006; NYC Economic Development Corporation 2006). "The model rule," said a coalition of environmental groups (Multiple Environmental NGOs 2006), "should reject anything which creates the impression that generators are entitled to allowances. We believe that 100 percent of the allowances should be allocated to consumers. ... [T]here will be significant windfalls to generators at the direct expense of consumers unless the predominant share of allowances is reserved for consumers." Other prominent comments also invoked the notion of public ownership explicitly and rejected any entitlement for emitters to any allowances. "First, the allowances are held in trust by regulators on behalf of the true, beneficial owners—the public," said the Pace Energy Law Project (2006), adding that "the public owns the right to pollute the commons" (see also Appalachian Mountain Club 2006). Others continued to invoke the polluter pays norm: "Generators should pay for the pollution associated with their products and recover it through market prices" (NY Energy Consumers Council 2006). At least sixteen other public comments from groups ranging from the Adirondack Council (2006) to the Catholic Diocese of Rochester, New York (2006) to 141 similar form letters from individuals echoed this call for allocating all allowances to consumers.

In addition, several groups noted the economic argument that generators would charge consumers for the cost of allowances even if given the allowances for free (e.g., Environmental Advocates of New York 2006). Some of these claims now invoked worries about apparent or potential windfalls for energy firms receiving free allocations of allowances in the EU ETS (e.g., MA Climate Action Network 2006; Clean Water Action 2006; NY Attorney General's Office 2006). By the time these references to the EU ETS entered the discussion, though, political momentum for auctioning allowances was well established.

Indeed, even as stakeholders were commenting on the draft model rule, several states had bills pending to auction 50 to 100 percent of their allowances. Vermont became the first state to require the sale of 100 percent of allowances on May 2, 2006. Concurrently, the New York Attorney General's Office spoke in favor of 100 percent auctioning at a May 2006 stakeholder meeting dedicated to the model rule—a moment that signaled to some advocates for the first time that "they could get all the allowances" (DeWitt 2011). In remarks on behalf of New York attorney general Eliot Spitzer, Jared Snyder emphasized the importance of the upcoming allocation decisions for states: "Allocation methodologies chosen now may well develop their own propulsion and provide the default choice for future reduction programs as well." For this reason, Snyder (2006) underscored the significance of 100 percent auctioning as a mechanism for preventing windfalls for electricity generators and to minimize RGGI's impact on consumers' electricity bills while offering an incentive "for development and implementation of clean sources of energy and energy efficiency." In this statement, the attorney general's office of a leading RGGI state reiterated the new issue frame that had been crafted and delivered by environmental policy entrepreneurs for the previous three years.

Advocates also drew on the new normative frame in debating which programs should qualify for the consumer benefit or strategic energy allocations outlined in the model rule. Environmental groups insisted that any consumer benefit allocations must "reduce the costs of the program for the state's ratepayers" while not replacing funding for existing programs or doing any harm to the environment (Environment Northeast 2006a; MA Climate Action Network 2006). In addition, environmentalists continued to focus on the idea of spending auction revenue primarily on energy efficiency subsidies rather than direct cash rebates as the best way to reduce the program's costs for consumers (ACEEE and Alliance to Save Energy 2006). Large energy consumers, by contrast, tended to favor rebates, arguing that existing energy efficiency programs were already well funded and that direct refunds of these revenues to ratepayers was a fairer option (Multiple Intervenors 2006; Consumer Power Advocates 2006). Arguments for direct rebates gained less traction with decision makers, however, in part based on the notion that investments in energy efficiency could protect consumers economically and reduce emissions—a double benefit that cash rebates would not offer.

Generators found themselves increasingly isolated as the old coalition against auctioning began to crumble in earnest. For example, large electricity consumers declared in May 2006 that they were "very concerned" about the cost impacts of RGGI and therefore supported 100 percent auction with rebates (RGGI Stakeholder Meeting Summary 2006, 6; Multiple Intervenors 2006; CT Industrial Energy Consumers Coalition 2006). This acceptance of auctioning by some large electricity consumers was another key turning point (Murrow 2010; DeWitt 2011). Meanwhile, National Grid (2006) reaffirmed its support for auctioning all allowances, and low-carbon energy producers also continued to support the auction idea (Entergy 2006), further isolating generators with higher rates of GHG emissions. Although other large industrial consumers of electricity remained opposed to auctions (e.g., Eastman Kodak 2006), these new defections were significant.

The public comments of generators opposed to auctions reflect the increasingly challenging political landscape. Despite the fact that they continued to express misgivings about auctioning, most remarks from large emitters were now focused on damage control in the form of limiting auctions to *no more than* 25 percent of all allowances (Dominion 2006; Conectiv Energy 2006; Edison Electric Institute 2006). The major objections offered remained the perceived threat to fuel diversity and reliability of electricity production. As American Electric Power's comment on the draft model rule stated, "First and foremost, the final Model Rule and state's implementing regulations should cap the size of the consumer benefit or strategic energy allocation to no more than 25 percent. To do otherwise would jeopardize the region's fuel diversity and electric system reliability." As confirmation of the problems facing generators at this point, American Electric Power (2006) also argued for a longer phase-in of any increase in auctions beyond 25 percent. Thus, even the strongest opponents of auctions appear to have seen the writing on the wall by May 2006, even before the final model rule was approved.

Only a few industry groups still challenged the new normative frame directly at this point, as when the Edison Electric Institute (2006) described auction proposals as "unfair" to coal- and oil-fired generators. Consistent with arguments in chapter 2 about the broad support for polluter pays and egalitarian norms, direct attacks on the new normative frame like this were uncommon, with those criticizing auctions being more likely to rely on the assertion that they were "untested," "unprecedented," or an "experiment" that deviated from accepted practices, and should be implemented gradually (Northeast Suppliers 2006; Northeast Regional GHG Coalition 2006; Independent Power Producers of New York 2006). As the Edison Electric Institute (2006) explained: "We strongly urge that the SO_x and NO_x models be the norm for allowance allocation ... and that the Draft [model rule] give direction to the states to distribute the allowances not by auction or other costly schemes, but equitably and without cost to the electric utilities, as has occurred under the Clean Air Act." Here is an excellent illustration of the new normative frame forcing generators to argue for the status quo simply because it is familiar rather than based on its economic or normative merits. Notice also that there is no mention of a norm of entitlement based on prior use or possession; at this point, the normative frame that supported the old model of cap and trade for many years had been fully replaced by the new public benefit frame.

At the same time, numerous energy producers and industry groups argued that the model rule was unfair to power producers with long-term pricing contracts, or those facing competition from electricity generators in states outside the RGGI cap (Independent Power Producers of New York 2006; American Electric Power 2006; New England Business Council et al. 2006; NRG Energy 2006). Some also contended that the resulting increase in emissions outside the cap would undermine the environmental benefits of the program, or sought to delay the model rule process until a new working group created to address these "leakage" concerns issued its report in December 2007 (e.g., NY Coalition of Energy and Business Groups 2006; Independent Energy Producers of New Jersey 2006; Northeast Suppliers 2006; Business Council of New York State 2006; Public Service Enterprise Group 2006).

Finally, in another indication of their growing acceptance of auctions, many power generators sought to give themselves first rights to bid on any auctioned allowances and increase the number of allowable offset credits

(Independent Power Producers of New York 2006; Keyspan 2006). Reflecting the influence of the new public benefit frame, generators often tried to justify these suggestions as in the best interest of consumers (e.g., Independent Power Producers of New York 2006). Generators also started to focus more on how auction funds should be spent, lobbying for consumer benefit allocations to promote energy efficiency or other ways to help ratepayers instead of going to a state's general fund (Dominion 2006).

After the close of the public comment period on the draft model rule, the SWG sponsored a stakeholder workshop in July 2006 on "implementing the minimum 25 percent public benefit allocation." The workshop organizers still avoided the word *auction*, preferring the term *public benefit allocation*, even though nearly all the presentations assumed the connection (e.g., Palmer 2006). Yet the workshop's title makes it clear that the 25 percent figure for public benefit allocation in the MOU was now seen as a floor on what states could do, indicating the continued momentum toward auctioning a larger percentage of allowances.

Several academic experts spoke at the July workshop, describing prior examples of government auctions of public resources such as the sale of spectrum rights for telecommunications in the 1990s and early 2000s as well as smaller experiments with auctions in Virginia under the NO_x budget program described in chapter 3 (Holt 2006; Gabel 2006; Shobe 2006). These presentations praised auctions for their ability to be "fast, fair, and generate high revenue" when designed properly (Holt 2006), while preventing "huge windfalls" to private firms that were formerly granted public resources at no charge (Kwerel 2006). Presenters also stressed the ability of auctions to supply much-needed revenue to state governments in a fair and efficient manner (e.g., Shobe 2006). In using the language of distributive fairness and windfalls, these presentations continued to reinforce the new public benefit framing even at a technical workshop ostensibly centered on efficient auction design. Power generator comments at the workshop were limited, and continued to fight a rearguard action in favor of minimizing the percentage of allowances to be auctioned and maintaining consistency across the RGGI states in how they organized their allocations (Santelli 2006; Savitt 2006).

After some delay based on the large volume of public comments, the SWG published the final model rule for RGGI on August 15, 2006. Although some important revisions were made to the offset rules, the core principles of the draft rule remained largely unchanged, particularly with regard to the allocation of allowances. The one substantive change in the model rule dealing with consumer benefit allocations was the new wording that a state will allocate "a minimum of twenty-five percent" of allowances to the consumer benefit or strategic energy accounts versus just "twenty-five percent" in the draft rule (RGGI 2007, sec XX-5.3[b]). This change is consistent with the prevailing political winds, indicating that states were already intending to auction a much larger portion of their allowances, if not all of them.

State Legislative Adoptions, 2006–2008

After the model rule was finalized, political support for 100 percent auctioning continued to build. By summer 2007, most of the participating states had committed in principle to auctioning the vast majority of their emissions allowances under RGGI. When the first RGGI auctions began in fall 2008, over 90 percent of all allowances were included, and auction revenues were designated for state energy efficiency programs as well as programs to lower energy costs for low-income households, consistent with the public benefit frame. Nor has the emphasis receded; RGGI (2013b) publications continue to promote auctions as a mechanism that helps lower carbon pollution, generate jobs, and reduce "family and business energy bills."

Although Vermont had already committed to auctioning in May 2006, advocates agreed that it was important to get a large state with substantial electricity generation capacity to make a similar commitment. A group of environmental advocates immediately started conversations with outgoing New York governor Pataki and his staff, lobbying for New York to enact regulations auctioning 100 percent of allowances before Pataki's term ended in December 2006 (DeWitt 2011). The NRDC played a central role in this effort in New York, while other environmental groups concentrated on other states (Bryk 2011; Litz 2011). The work apparently paid off, as New York issued the first draft of its rules for auctioning nearly all allowances in December 2006 ("New York 100% Auction Plan" 2006). Meanwhile, new Massachusetts governor Patrick announced his intention to rejoin RGGI and auction all allowances on January 18, 2007 ("Patrick Signs Up Bay State" 2007). Thus, by January 2007, the governors of the region's two largest emitting states had declared that they intended to auction all or nearly all of their allowances.

Interviews with elected officials and their senior staff in New York and Massachusetts confirm the political importance of using auctions to lower RGGI's impact on consumers. As Governor Pataki (2014) later recalled, protecting consumers was a vital reason for his support for RGGI's auction provisions. "If you want to get serious about this," he recalled saying at the time to environmental advocates such as the NRDC, "you have to understand the difficult economic times and that the program has to not put new burdens on consumers or industry." Other New York State officials remember similar arguments resonating with the governor (Litz 2011). Media articles covering the release of the December 2006 rules also quote environmental advocates saying "allowances are a public good, and the public owns the right to pollute" as well as state administrators arguing that auctions will benefit consumers directly "by promoting important environmental and energy-saving investments" ("Impact of Pollution Plan Debated" 2006; Nearing 2006).

Meanwhile, in Massachusetts, former senior staff members for Governor Patrick also confirmed that dedicating auction revenue to energy efficiency programs in order to protect ratepayers from higher energy costs was vital to the governor's decision to approve auctions (Bowles 2014; Cash 2014). Indeed, this recommendation was informed in part by an earlier analysis by the Romney administration of the economic benefits of returning auction revenue to consumers (Bowles 2014; Cash 2014). As a former undersecretary for policy in Patrick's Department of Environmental Protection (DEP) recalled, the new governor and his staff realized that "these permits have value. Let's have the value of those permits got to ratepayers" (Cash 2014).

In this respect, it appeared that when elected officials "got" the logic behind auctions, they tended to ask, Why not auction all of them? (Kaplan 2011; Litz 2011). As one leader in the SWG recalled, even Governor Romney remarked in 2005 that "if you're going to have a market, have a real market and auction all of the allowances" (Hamel 2013). Another senior SWG member remembered Governor Pataki making a similar argument at a meeting in 2006: "I can understand 100 percent, or 0 percent, but not 25 percent" (Litz 2013).

Soon after the Massachusetts announcement, Rhode Island declared its intention to rejoin RGGI. This brought the total number of RGGI states up to ten, with Maryland's decision to join in April 2006. Meanwhile, Maine officials announced that they favored auctioning all allowances in February

2007, while Connecticut governor Jodi Rell stated that she favored selling all allowances in March ("Editorial" 2007; "Connecticut Governor" 2007). Just seven months after the finalization of the model rule, a majority of RGGI states had publicly committed to auctioning all or nearly all allowances, with the others soon to follow. This outcome was consistent with earlier media reports indicating that most states were inclined to follow New York's move in 2006 toward auctions ("Other States" 2006). As climate change policy expert David Doniger put it in April 2007, the idea of the free distribution of GHG allowances was now "teetering and almost dead" (cited in "EU Experience" 2007).

Although New York adopted auctions through administrative action, other RGGI states needed legislation to implement their governors' recommendations for auctions in enacting RGGI (table 4.3). In lobbying state politicians regarding their allocation decisions, environmental groups continued to use the public benefit frame in tandem with economic arguments against free allocation (Murrow 2013; Stratton 2014). One environmental advocate involved in state legislative lobbying later summed up the process by declaring "we reframed the question" by describing auctions as a mechanism for addressing higher consumer costs from the emissions cap (Kaplan 2014).

An Environment Northeast policy brief distributed to Connecticut lawmakers, for example, began by arguing, "Air quality and the world's climate are a public good that polluters do not have the right to spoil." The same document estimated the substantial monetary value of RGGI allowances for different states, and cited statements by National Grid and large electricity consumers in asserting that the consumer benefit allocation should be used to "reduce the costs of the RGGI program to the state's electricity ratepayers" (Environment Northeast 2006b). Media reports picked up on and echoed these same frames, such as a *New York Times* article on RGGI with the headline "States Aim to Cut Gases by Making Polluters Pay" (Barringer and Galbraith 2008) As with the public comments on the model rule, references to the apparent windfall for some EU emitters from free allowance allocation also appear in news articles covering the state legislative process (e.g., "Impact of Pollution Plan Debated" 2006; Nearing 2006; Miller 2007). In this respect, the potential for generating consumer benefits from auctions was a central part of these legislative deliberations about adopting the program (Kaplan 2014; Murrow 2013, Stratton 2014).

Table 4.3
RGGI State Auction Legislation/Regulations

State	Date	Bill and regulation number	Allowances auctioned	Revenue to energy efficiency and low-carbon energy	Revenue to ratepayer rebates
Vermont	5/2/06	H.860	100%	100%	None
Maine	6/1/07	HP 1290	100%	All revenue from prices < $5/ton	All revenue from prices > $5/ton
Connecticut	6/4/07	Act 07-242	Up to 100%	92.5%	None
Rhode Island	7/2/07	H5577	100%	Undetermined by statute	Undetermined by rule
New Jersey	1/13/08	HB 4559	Up to 100%	Up to 80%	None[a]
Maryland	4/24/08	SB 268	85%	56.5%	40%
New Hampshire	6/11/08	HB 1434	99%	All revenue from prices < $6/ton	All revenue from prices > $6/ton
Delaware	6/30/08	SB 263	60% (2009) 100% (2014)	65% to 80%	Up to 15%
Massachusetts	7/2/08	SB 2768	99%	100%[b]	None
New York	9/8/08	Chap. 3, sec. 242-5	~ 95%	100%[c]	None

[a] If price > $7/ton (two consecutive auctions), must develop plan for "immediate ratepayer relief."

[b] Except limited reimbursements to local communities for lost property taxes from devalued or decommissioned power generation facilities.

[c] Except funds dedicated to program administration.

This reframing changed the primary political topic in many cases from whether or not to auction, to what to do with the revenue. Indeed, one agency head stated that the biggest issue in discussing auctions with governors and their staffs was deciding where the revenue should go rather than whether to have an auction or not (Sheehan 2012). On this point, it appears that most elected officials were more persuaded by the promise of a "double dividend" from investing auction revenue in energy efficiency programs, thereby saving consumers money and increasing local economic development related to green technology (Kaplan 2014; Pacheco 2014). Although some governors and legislators wanted a direct rebate to consumers, especially if allowance prices went over a minimum level (Stratton 2014; Lord 2008), most elected officials seem to have been more interested in the additional environmental and economic benefits of investing in state energy efficiency or renewable energy programs, as confirmed by the pattern of spending mandates by different states in table 4.3.

After the first wave of RGGI states enacted legislation approving nearly 100 percent auctioning, attention turned to New Jersey, where the conflict over RGGI was particularly acute. New Jersey was another early supporter of RGGI, and had a unique history dating back to the 1990s of taking action on climate change (Rabe 2004). In addition, more than one interviewee noted that New Jersey officials were also leaders on the idea of auctioning allowances (Litz 2011; Bryk 2014).

Despite this long-standing commitment to ambitious state climate policy, however, New Jersey was in a difficult position with regard to RGGI. Unlike many RGGI states, New Jersey is part of an electricity network connected to Pennsylvania and other states outside the RGGI cap. As a result, New Jersey electricity generators were more vulnerable to competition from power suppliers that did not have to reduce their emissions. This threat of cheaper electricity imports from out-of-state generators made ratification of New Jersey's participation in RGGI, including the decision to auction allowances, particularly contentious. The state eventually enacted the Global Warming Solutions Fund Act on January 13, 2008, but only after more than a month of intense political maneuvering and negotiations, including attempts to limit how much power companies would have to pay for allowances or to exempt certain types of generators from the auction process ("New Jersey Legislature" 2008). As one environmentalist commented on a version of the act just prior to the final vote in both houses, "The loopholes in this bill allow some polluters to pay less, some polluters to pay nothing at all, and some polluters to actually get paid" (quoted in Nussbaum 2008). Even in their protestations against efforts to weaken New Jersey's commitment to auctions, environmentalists continued to rely on the public benefit model, and a large majority (80 percent) of the state's auction revenues were dedicated to energy efficiency programs benefiting ratepayers.

When Massachusetts enacted its Green Communities Act on July 2, 2008, all the RGGI states had committed to auctioning a majority of their allowances, with most states effectively auctioning all of them (table 4.3). Moreover, all states except Maryland dedicated most auction revenue to energy efficiency programs to help consumers rather than to direct ratepayer rebates. When New York issued its final regulations on September 8, 2008, ratifying its commitment to auction approximately 95 percent of its allowances, the initial fight over auctions in RGGI was over. On September 25, 2008, the newly created umbrella organization RGGI Inc. auctioned approximately 12.5 million allowances for six of the ten RGGI (2015c) states, at a clearing price of $3.07 per ton, raising approximately $38.5 million.

Over the next six years, RGGI (ibid.) conducted a total of twenty-nine auctions, selling more than 785 million allowances and raising more than $2.2 billion in revenue. Of that revenue, approximately 70 percent was dedicated to consumer energy efficiency or clean energy programs, and 15 percent to direct ratepayer relief, with the remainder going to administration or other programs to reduce GHG emissions (RGGI 2015d). Three states also briefly diverted RGGI auction funds toward deficit reduction or state general fund expenses, creating renewed controversy over the appropriate definition of public benefit that was so central to the creation of the auction idea in the first place. Chapter 5 considers these ongoing debates over the disposition of auction revenues in other programs in more detail.

Conclusion

When asked in retrospect what the key factors were in the decision to auction allowances under RGGI, SWG cochair Sonia Hamel (2010) cited three things:

1) The argument that utilities would charge consumers for allowances either way
2) The argument against "giving away something the public already owns"
3) The idea that RGGI could meet its goals more cheaply by auctioning allowances and investing the revenue in energy efficiency

Environmental advocates echoed this assessment during the allocation debate: "To insure this gets broad support, we need to create a program that produces a tangible benefit. We need to show that this program addresses the very serious problem of global warming and pollution but does it in a way that produces a benefit people can get their arms around" (RGGI Stakeholder Meeting Summary 2005c, 29).

Elected officials offered similar explanations. As noted earlier in the chapter, former New York governor Pataki (2014) recalled discussing the issue of auctions in RGGI in the following manner: "If you want to get serious about this, you have to understand the difficult economic times, and that the program has to not put new burdens on consumers or industry." And those representing industry agreed that the dedication of auction revenue to public energy efficiency programs was an important part of the proposal's political strength (Bradley 2011; Svenson 2011). As a leading representative of industrial interests concluded, a key argument was that "if ultimately ratepayers are going to pay for this in some fashion, they should get the benefit" (Ruddock 2012). In addition, industry representatives credited environmental groups for their commitment to the auctioning concept and ability to influence regulators on the issue with effective arguments during RGGI's design phase (Svenson 2011; Ruddock 2012).

All four perspectives are consistent with the significance of connecting auctions to the idea of protecting electricity consumers from higher energy prices, as suggested by the new public benefit framing of the issue. Thus, a detailed review of the process by which RGGI states elected to auction nearly all emissions allowances supports the normative reframing explanation outlined in table 4.2, including the critical role of the new public benefit frame. First, this chapter showed how a small band of environmental advocates were instrumental in putting the idea of auctioning on the RGGI policy design agenda, despite the initial misgivings of state agency officials. While elected officials made the final choice to auction, previous explanations have overlooked the important groundwork initially laid by environmental advocates, and then later by agency staff. In this respect, the chapter has illustrated how auctions were a part of the RGGI policy discussion from the beginnings of the process in 2003 as opposed to entering the process at a later date.

The chapter has also shown how the existence of state-run energy efficiency programs in these states, funded by public benefit charges enacted in the 1990s, was crucial to making elected officials more comfortable with the idea of auctioning. In the same vein, it is clear that auction revenue was politically important as a device to mitigate rate increases for most of the public rather than to divide industrial and business opposition.

Finally, the chapter has described the prominence of the new public benefit normative framing about allowance auctions throughout the entire design process. Although arguments about economic windfalls for power generators from free allowances were critical, they were inadequate to generate a move toward auctioning in the similar process of designing the NO_x budget programs. The normative frame and policy design in RGGI that also incorporated a distribution of broad public benefits from any use of the atmospheric commons left generators struggling for an effective political response, and facilitated a complete reversal of the old model of emissions trading that presumed allowances would be given away for free to existing users.

As such, it is no overstatement to say that RGGI really did constitute a policy revolution—the emergence of a startling new paradigm of public ownership. Advocates went into the process asking for 100 percent auctioning, but thinking it would be "incredible" if they could get even 25 percent (Murrow 2010). Europeans involved with the EU cap-and-trade program being developed in this period appear to have been surprised that RGGI's designers were able to sell nearly 100 percent of their allowances, whereas the European Union had been able to auction less than 5 percent of its allowances in the first phase of its ETS in 2005–2007 (Burtraw 2011; Hamel 2010).

While some have maintained that RGGI's adoption of auctions was less significant due to the relatively low price of allowances in the early stages of the program and the relatively modest emissions reductions required, this argument misses a vital point. The idea that polluters would pay *any* price for their emissions was a political nonstarter, despite decades of being supported by economists, until RGGI. By changing the norms applying to allocation, RGGI created a "new normal" where polluters are expected to pay, with limited exceptions, for their use of the atmospheric commons. The new default assumption about entitlement to the atmosphere has shifted, much as the New York Attorney General's Office predicted back in 2006, not just in RGGI, but in other programs, as will be discussed further in chapter 5.

5   Other Applications of the Public Benefit Model and Normative Reframing

Even as RGGI held its first allowance auction, a backlash to cap and trade was gaining strength. For a variety of reasons, partisan opposition to *any* government action on climate change intensified in the United States and other nations starting in 2008. After the defeat of a leading federal emissions trading proposal to address climate change in the United States in 2010, some began to declare cap and trade "dead" as a policy option. This chapter rejects the asserted demise of cap and trade, arguing instead that the public benefit model for climate policies created in RGGI offers the best hope for political progress on the problem of climate change even in the face of stronger opposition to any climate policy. The chapter also contends that consistent with public benefit framing, the key question for current and future climate policies is becoming who should benefit from the private use of the atmospheric commons in terms of the disposition of allowance auction revenues or other benefits.

The chapter begins with a review of post-2008 climate policies reaffirming the principles of public benefit framing. This survey indicates that many leading climate policies and proposals now use cap and trade with auction, or other methods of charging polluters for their emissions, consistent with the new framing introduced in RGGI. In the case of RGGI itself, for example, the program's defenders resisted new political threats from Republican critics by reaffirming the broad public benefits of the energy efficiency programs supported by auction revenue. In addition, two other leading climate policies worldwide have followed much of the public benefit model in creating a substantial role for auctioning allowances: the EU ETS, and California's cap-and-trade system implemented under the state's Global Warming Solutions Act. The recent history of failed climate change policy efforts also suggests the relevance of the public benefit frame. Consideration of post-2008 federal cap-and-trade proposals in the United States and Australia, for instance, suggests that deviations from the public benefit model contributed to the ultimate failure of both efforts.

In all these cases, the question of how to allocate revenue from the sale of emissions allowances remains a leading political issue. Climate policy advocates are now promoting variations on the public benefit frame emphasizing reduced climate change impacts or even health improvements from reducing copollutants associated with GHGs. How persuasive these different variations on the public benefit frame will be is a key future question for climate policy—one that is already being tested in California and in policy deliberations over the 2015 US EPA rules to reduce GHG emissions from power plants. Indeed, it is arguable that the disposition of benefits from the use of the atmospheric commons is poised to overtake long-standing conflicts concerning the science of climate change as the most politically salient issue in these policy debates.

The chapter also explores the applicability of public benefit framing to other policies beyond cap and trade. Policies that use different strategies for pricing carbon emissions such as carbon taxes become good candidates for public benefit framing by dedicating that tax revenue directly to consumer benefits, as appears to have already happened in one prominent case in British Columbia. Other policy mechanisms governing the private use of the atmospheric commons could draw on public benefit framing to the degree they require polluters to bear the costs of using this public resource and generate potential benefits for a wide range of society. Moreover, the public benefit frame could easily be applied to resource management and pollution conflicts related to other common resources, such as water pollution, public lands management, or minerals development.

Finally, the chapter moves beyond emissions trading to explore the wider implications of the normative framing model of policy change. If the model of norm-driven policy change is correct, it should be possible to predict future policy change (or stability) for other issues by evaluating which policy designs exhibit the best "normative fit" with the new frame in the eyes of the public. The chapter therefore concludes with a discussion of how scholars and change advocates might use normative reframing concepts to identify other policies that are ripe for change using such a strategy.

The Expansion of Cap and Trade with Auction Since 2008

Despite a few prominent setbacks, cap and trade remains the dominant policy for addressing climate change worldwide, with new programs springing up in Quebec, South Korea, China, Japan, and New Zealand, as well as California's major cap-and-trade experiment (Newell, Pizer, and Raimi 2013; Betsill and Hoffmann 2011). As of 2015, the largest of the world's cap-and-trade programs for GHGs all auctioned a majority of their allowances, including the EU ETS, RGGI, California, Quebec, and Switzerland (International Carbon Action Partnership 2015), in some cases creating large sources of public revenue. In fact, programs relying primarily on auctions represented approximately 56 percent of the world's total capped GHG emissions as of July 1, 2015, while programs with no provisions for auctions regulated only 11 percent of those capped emissions (table 5.1). Few of these programs existed ten years earlier, and those that did auctioned only a tiny percentage of their allowances. In this respect, the early evidence suggests that the public benefit model has already started to transform the

Table 5.1
Global Distribution of GHG Cap-and-Trade Programs by Auction Provisions, July 1, 2015

Primarily auction (more than 50% of allowances)		Some auction (less than 50% of allowances)		No auction	
Program	Annual cap*	Program	Annual cap	Program	Annual cap
RGGI	91	South Korea	573[b]	Japan Saitama	11
California	395	PRC Shenzen	32	Japan Tokyo	14
EU ETS	2,040[a]	PRC Hubei	324	Kazakhstan	153
Quebec	65	PRC Guangdong	408	PRC Beijing	50
Switzerland	6	PRC Shanghai	160	PRC Chongqing	125
				PRC Tianjin	160
Total allowances	2,597		1,497		513
% allowances	56%		33%		11%

* In mega tonnes CO_2 equivalent.

[a] Stationary sources only.

[b] Auctions scheduled to start in 2018.

Source: International Carbon Action Partnership 2015.

political landscape for climate policy, with cap and trade becoming a leading option for controlling GHG emissions, and the largest of such policies now relying on auctions as their primary allocation method.

An analysis of expenditures of revenue from leading cap and trade with auction or carbon tax policies also helps illustrate the influence of the public benefit model. Through 2013, more than two-thirds of all revenue from the auction of GHG allowances or carbon taxes in RGGI, California, the EU ETS, British Columbia, and Alberta went to energy efficiency programs, tax credits, or direct dividend payments designed to directly lessen the economic impact of the program on all consumers (Sekar, Munnings, and Burtraw 2014). By comparison, 25 percent of these revenues were spent on low-carbon technology research and development, and only 7 percent was dedicated to other government programs (ibid.). This data also suggests how influential the public benefit frame has been, given how little money has been dedicated to programs that do not either directly benefit consumers or directly address the problem of climate change, consistent with the RGGI model. At the same time, they also indicate the growing tension between a "narrower" vision of public benefits promoted in RGGI requiring a distribution of auction or tax revenue to all or nearly all consumers on a roughly equal basis, and a "broader" vision allowing for a wider range of programs like research on low-carbon technologies that offer more indirect benefits to citizens.

Although the dramatic spread of auctions and similar pricing mechanisms does not by itself confirm the influence of the public benefit model, it is a noteworthy development that is sometimes overlooked in policy discussions about the decline of cap and trade. The dedication of most revenue to programs to help consumers is especially noteworthy in this regard. For this reason, conversations about the future of climate policy must start from this fundamental empirical point: cap and trade is the most common policy for regulating GHG emissions in the world today, the substantial majority of those allowances are now auctioned rather than given away, and much of that revenue is dedicated to helping energy consumers.

Public Benefit Framing in the Expansion of Cap and Trade with Auction

A deeper look at the expansion of cap and trade with auction in three cases confirms the growing relevance of the public benefit model. During a period of economic recession in Europe and the United States, and intensified partisan opposition to any policies threatening to raise energy prices and hurt the working class, RGGI and the EU ETS still managed to strengthen their emissions reduction goals. Defenders of both policies accomplished this goal by stressing the advantages of the public benefit model using cap and trade with auction, which they began to describe as a "cap-and-invest" strategy (Tanzler and Steuwer 2009). Public benefit arguments also underwrote the implementation of a major new cap-and-auction program in California in 2010, although in California political controversy has been more intense over what to do with auction revenue. In this sense, post-2008 developments in RGGI, the EU ETS, and California highlight the expanding influence of the public benefit model as well as the growing importance of the question of how to spend auction revenue.

Defending RGGI

RGGI states faced the same factors that made climate change policy more difficult in the United States and other nations after 2008: a newly energized conservative political movement hostile to climate change policies, and a global economic recession that made it hard to promote any policy that might further burden the economy or consumers. Consistent with these trends, New Jersey, Delaware, New Hampshire, and Maine all experienced organized efforts to compel a withdrawal from RGGI. With the departure of several governors who made the initial commitment to RGGI after 2008, the program was even more vulnerable to these new political threats.

No effort to withdraw from RGGI succeeded, however, except in New Jersey, where Governor Chris Christie left the program in 2011 without legislative approval. Public benefit framing was at the heart of the successful efforts to prevent repeal of the program in other states. New political leaders, such as New Hampshire governor John Lynch, defended the program using the normative framing that made RGGI politically viable in the first place. "These are funds," said Governor Lynch of the threatened loss of auction revenues, "that have been invested directly in helping New Hampshire families, businesses and local governments become more energy efficient, reduce costs and create jobs" (quoted in Spolar 2011).

Another attack on the public benefit model arrived when state budget crises led several RGGI states to dedicate allowance revenues to their general funds rather than to consumer benefit programs. The growing political significance of using auction revenue for energy efficiency programs to benefit the public is indicated by RGGI supporters' condemnation of these proposals as "RGGI-cide" (Marshall 2010). Consistent with the new model of cap and trade that made RGGI's auctions possible, most of these diversion efforts failed, and the vast majority of RGGI auction revenue continues to go toward programs that directly benefit energy consumers (see table 5.2).

A related threat to RGGI's public benefit model came, ironically, from lower allowance prices due to reduced energy demand from the economic recession. This development challenged the integrity of RGGI by reducing the funds available for public benefit programs. Fortunately, RGGI's auction designers incorporated a minimum "reserve" price in their auctions that helped stabilize the allowance market and maintain a base level of revenue for public benefit programs that were crucial to political support for the new RGGI model. The lack of a similar reserve price created political difficulties in other programs relying on the public benefit model, such as the EU ETS.

Table 5.2
Allocation of RGGI Auction Revenue through 2013, by State

State	Energy efficiency	Consumer clean energy	Direct bill assistance	GHG abatement	Admin/ other
Maine	96%	0%	0%	0%	4%
Massachusetts	93%	0%	0%	5%	3%
Rhode Island	89%	0%	0%	0%	11%
Vermont	98%	0%	0%	0%	2%
Delaware	61%	11%	10%	10%	9%
Maryland	25%	6%	59%	4%	5%
New Hampshire	78%	0%	19%	0%	4%
Connecticut	72%	20%	0%	0%	9%
New York	56%	16%	0%	20%	8%
RGGI total	**62%**	**8%**	**15%**	**9%**	**6%**

Note: Percentages for each row may not equal 100 percent due to rounding.

Source: RGGI 2015d.

In 2012, RGGI underwent a required program review in order to set a new emissions cap and consider possible modifications to the program. In preparation for this review, RGGI commissioned a study of the economic impacts of the investment of auction revenue in the participating states. The study (Hibbard et al. 2011), which received substantial media attention (e.g., Gallucci 2012b), concluded that auction revenue had resulted in major reductions to consumers' electricity bills as well as net economic benefits to the participating states (see figure 5.1). Subsequent reports for the program review continued to focus on the positive consumer and economic benefits of auction revenue investments, noting that 66 percent of all auction revenue had been invested in energy efficiency programs, and another 17 percent was dedicated to assisting low-income consumers directly with their bills (RGGI 2012). These reports particularly emphasized the benefits of auction revenue for average individuals, as exemplified by this quotation from a Vermont homeowner: "After Tropical Storm Irene flooded our house, we resolved to rebuild better and smarter. Efficiency Vermont [a public benefit program funded by RGGI auction revenue] was there to help us through the whole process—and now our home is safer and more comfortable. And, best of all, we save money on our energy bills every month" (quoted in RGGI 2012).

Consistent with this study, the program review sought to "ensure the continued success" of RGGI's (2013c) approach to "effectively reducing CO_2 emissions while providing benefits to consumers and the region." The review remained committed to what RGGI (2013a, 2013c) leaders were also now calling an "auction-and-invest" strategy while aggressively lowering the emissions cap by 45 percent in 2014, and 2.5 percent each subsequent year through 2020. In addition, the program review recommended new measures to reduce the backlog of unused allowances from the first five years of the program, further strengthening the allowance market (RGGI 2013c). These dramatic reductions of the RGGI cap after a few years of the program's operation run counter to the claims of some critics that the program has been largely symbolic, with only token emissions reductions.

By 2014, all the RGGI (2014b, 2015b) states had ratified these new rules, and auctions were already generating higher allowance prices (more than $5 per ton) based on the more stringent cap. As in the program review, RGGI press releases highlighted how auctions created funds for investments in energy efficiency and renewable energy (RGGI 2014a, 2014b).

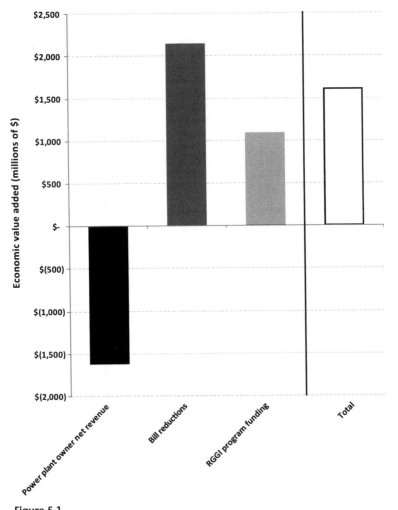

Net economic impact to states in the RGGI region (2011$)

Figure 5.1
The economic impact of RGGI auction revenue investments on the RGGI region.
Note: figures represent dollars discounted to 2011 using a 3% public discount rate.
Source: Hibbard et al. 2011.

By this point, RGGI's energy efficiency programs had supported a cumulative total of 1.2 million households in the region, while direct bill assistance had helped more than 2.4 million households (RGGI 2015d). In this sense, RGGI advocates became even more dedicated to their public benefit normative framing as they defended and strengthened the policy under more challenging political conditions. And their strategy appears to have worked in the short term, as other states publicly considered joining the RGGI partnership and the EPA began citing RGGI as a model for compliance with its new federal regulations for reducing GHG emissions (Ress 2014; Holden 2014).

Although RGGI states remain committed to the narrower vision of limiting auction revenue to programs with direct consumer benefits, they do vary in their relative commitment to energy efficiency versus other programs (table 5.2). One group of states—Maine, Massachusetts, Rhode Island, and Vermont—has invested nearly all auction revenue into energy efficiency. Most of these programs focus on homeowners and individual ratepayers, except in Rhode Island, where a larger percentage is dedicated to energy efficiency programs for businesses.

A second group of RGGI states has included provisions for direct ratepayer assistance. Although cash assistance for electricity bills was not something emphasized by most environmentalists in their version of public benefit framing, it became a leading part of the RGGI design in Maryland in particular, and plays an important role in Delaware and New Hampshire. In both Maryland and Delaware, this bill assistance is mostly targeted to low- and moderate-income ratepayers. In New Hampshire, however, a 2012 law redirected a large portion of future auction revenue to equal rebates for all ratepayers regardless of income (Backus 2015).

Finally, a third group of states has dedicated more auction revenue to programs supporting the adoption of new clean energy sources (identified as "consumer clean energy" in table 5.2), again in a way designed to reduce electricity demand and economic impacts on consumers. These states are led by Connecticut and New York, which invested between 16 and 20 percent of their auction revenue through 2013 in programs to subsidize the purchase and installation of renewable energy systems by state residents. Although New York's focus is on the residential adoption of these technologies, in Connecticut there is more support for energy efficiency programs for businesses, including RGGI funding for a "Green Bank" to help provide lower-cost financing for clean energy and energy efficiency projects at commercial and municipal facilities.

New York has also invested the most of any RGGI state in programs that reduce GHG emissions without necessarily reducing electricity bills for consumers (the "GHG abatement" category in table 5.2), moving furthest from the narrower vision of the public benefit frame requiring ratepayer relief. For example, New York has invested RGGI funds in "sustainability planning" for local communities that seek to reduce emissions from the transit sector as well as create more energy efficient buildings. A few other RGGI states have made similar investments, including Delaware's use of auction revenue for a broad range of GHG abatement and climate adaptation programs. In expanding the scope of expenditures from auction revenue in this manner, these states are part of the larger debate over the appropriate use of revenue from a public resource like the atmospheric commons.

In sum, this comparison of spending by RGGI states represents two important conflicts in the disposition of auction revenue. First, all RGGI states remain strongly committed to spending the majority of their auction revenue on programs to benefit consumers directly by reducing energy demand or providing direct bill assistance, consistent with the original vision of public benefit framing. Two RGGI states, though, now dedicate a large percentage of their auction revenue to direct bill assistance, as New Hampshire decided in 2013 to return most auction revenue directly to ratepayers in future years. The shift toward direct bill assistance in New Hampshire in 2013 is noteworthy, and represents another crucial conflict dating back to Barnes's (2001) sky trust idea: Should public benefit funds be used to reduce consumer energy consumption and therefore expenses, or be returned directly to citizens on a per capita basis?

Second, several RGGI states have started to invest auction revenue in a wider range of programs that reduce GHG emissions or promote climate change adaptation, but do not necessarily reduce the economic burdens of citizens. New York is especially significant in this regard, given the relative size of its auction revenues and increasingly diverse range of emissions reduction programs being funded. This diversification of investments from auction revenue into other programs to reduce GHG emissions introduces the key new climate policy conflict created by public benefit framing: Are government programs designed to mitigate climate change more broadly (and meet other environmental or social justice goals) still consistent with the public benefit framing underwriting the cap-and-invest strategy? Similar conflicts over how to spend auction revenue have been even more intense in other programs, and threaten to be a crucial issue for climate change policies going forward.

Extending the EU ETS

Controversy over cap and trade was not limited to the United States. In the European Union, problems with allowance price volatility, renewed political opposition led by large GHG emitters concentrated in Eastern Europe, and charges of fraud bedeviled the EU ETS as it moved beyond its pilot phase in 2008. Despite these challenges, the program strengthened its emissions reduction goals after 2008 through greater reliance on public benefit framing along with associated policy designs for expanding auctions and investing the revenue toward broad public benefits.

As mentioned in the previous chapter, allocation design flaws led to several problems for the EU ETS in its pilot phase. As the pilot phase neared completion in 2007, allowance prices collapsed due to an oversupply in the market and the inability to use these allowances in future years (Ellerman et al. 2010, 142). The price collapse gave the program a black eye, and resulted in revisions to ETS rules to limit the overallocation of allowances in the future and allow firms to "bank" unused allowances for use in future compliance periods to reduce price volatility. In addition, many argued that free allocations to the power sector in the pilot phase had resulted in windfall profits for many firms, much as free allocations were expected to have done in RGGI (Matisoff 2010).

In the face of these early problems, EU member states and the European Commission recommended several important reforms during a 2007 program review, including:

1) Replace member state emissions caps with a single, EU-wide cap to reduce the risk of flooding the allowance market with "cheap credits" created by nations that set their caps at unrealistically high levels

2) Prevent windfall profits by auctioning most allowances, and dedicating that revenue to programs to help mitigate and adapt to climate change as well as to subsidizing compliance efforts by lower-income nations within the European Union (Skjaerseth and Wettestad 2010, 106–107)

The December 2008 directive extending the EU ETS largely reflected these goals, doing away with national allocation plans in favor of an EU-wide cap that reduces emissions by 20 percent from 2005 levels by 2020, and requiring far more allowance auctioning (Convery and Redmond 2013). Starting in 2013, the rules required electricity generators to purchase all their allowances at auction, with other industries having to buy at least 20 percent of their allowances by 2013, rising to 70 percent by 2020 (Matisoff 2010).

Important exceptions provided free allowances for large electricity generators in certain eastern bloc nations highly dependent on coal, to be gradually phased out by 2020, as well as for energy-intensive industries facing
global competition from firms in uncapped countries (Skjaerseth and Wettestad 2010, 106), in order to facilitate their entrance into the program.

In this sense, the post-2013 EU ETS looked a lot more like RGGI, rejecting
the old model of cap and trade for a program in which emitters are expected
to pay for their use of the atmospheric commons unless they can argue for
a limited number of free allowances as "compensation" for high transition
costs or competition from uncapped sources. In contrast to RGGI, however,
the EU program adopted a broader notion of public benefit that dedicates
auction revenue to a larger range of climate change mitigation and adaptation efforts, rather than primarily to consumer protections or rebates. Finally,
the EU plan for 2013–2020 also does not require member states to use auction revenue for climate change mitigation and adaptation programs, relying instead on a nonbinding provision recommending that states dedicate at
least 50 percent of such revenue to those purposes (ibid., 107).

Unfortunately for EU ETS supporters, these revisions failed to accom-
plish the primary goal of stabilizing allowance prices. As in the 2005–2007
pilot phase, the EU ETS continued to be marked by an oversupply of emissions allowances due to depressed economic activity from the global recession of 2008–2009, a steady flow of cheap allowances from international
offset projects, and additional requirements to generate energy from renewable sources, thereby limiting demand for allowances (Convery and Redmond 2013; Stavins 2012). This oversupply of allowances led in turn to a
major decline in EU ETS allowance prices, from nearly thirty euros per ton
in 2008 to less than five euros in 2013. In addition, more than one highly
publicized example of fraudulent trading or electronic theft of allowances
hurt the reputation of the program (Chaffin 2012; Kantner 2011).

These problems threatened the ETS in several ways. Low allowance
prices reduced the incentive for firms to make costly investments to reduce
their emissions, slowing the pace of technological change required to meet
more ambitious future emissions reduction goals. Low allowance prices also
reduced auction revenue available to fund public GHG mitigation efforts,
including work on carbon capture and storage, renewable energy, and
energy efficiency programs (Convery and Redmond 2013). Finally, public

scandals about fraud reduced confidence in the integrity of allowances as secure assets, threatening a core requirement of any successful emissions trading program. More generally, the problems increased pessimism about the ETS overall, as reflected in media reports describing the policy as offering "cheap and dirty" allowances (Chaffin 2012), and amounting to a failed experiment that had done little to address climate change (Carrington 2013; Reed and Scott 2013).

In response, the European Commission continued to propose reforms. Allowance tracking and electronic security were tightened up to prevent future trading scandals (Bierbower 2011). More controversially, the commission adopted the idea of "backloading" allowances, or reducing the emissions cap by moving some allowances to later years of the program (Harvey 2013). This proposal was designed to reduce the current supply of allowances, thereby increasing the allowance price. Other suggestions for reform, such as permanently eliminating some excess allowances or creating a reserve price for allowance auctions as was done in RGGI, had yet to be adopted as of 2014, but remained part of the policy conversation (Hope 2014).

Thus, like other climate policy efforts, the EU ETS faced a difficult five years from 2008 to 2013. Despite these challenges, however, the European Union adopted new GHG reduction goals for 2030, with the ETS remaining a central mechanism for compliance (Neslen 2014). Many proposals to reform the EU system have emulated the RGGI model, such as increasing reliance on auctions as well as recommendations to add an auction reserve price to maintain a more consistent economic incentive for investment in new carbon mitigation strategies and technologies (Convery and Redmond 2013; Grubb 2012; Dirix et al. 2013). In addition, EU members have dedicated 87 percent of their auction revenues through 2013 to programs "for climate and energy related purposes," well above the nonbinding 50 percent threshold recommended in the program renewal directive (European Commission 2014, 18).

In this manner, EU ETS designers have imitated RGGI in important ways as they struggle to improve their program in the face of resistance from high-emitting industries, especially in coal-dependent countries led by Poland and other eastern bloc nations. Although the final outcome is uncertain, changes consistent with the public benefit frame have already helped extend the political life of the EU ETS.

Establishing a New Cap-and-Auction Program in California

Even as the RGGI states had finished negotiations over their MOU and were beginning work on their model rule, California was in the midst of major political changes that would lead to the adoption of the most ambitious GHG reduction goals in the United States to date. Representing a compromise between newly elected Republican governor Arnold Schwarzenegger and a Democratic-controlled state legislature, the 2006 Global Warming Solutions Act (AB 32) committed the state to reducing its GHG emissions to 1990 levels by 2020, and to 80 percent below 1990 levels by 2050. The law also required the California Air Resources Board (CARB) to create a plan for meeting this emissions reduction goal, stating that the board "may" use emissions trading as one way to meet the new emissions target (Rabe 2013; Farrell and Hanemann 2009). As a result, California has used a variety of policies to meet the overall emissions reduction goals in AB 32, including a statewide cap-and-trade program that covers the full "basket" of GHGs rather than just CO_2, and incorporates emissions from both the transportation and electricity sectors (C2ES 2014). Although it only affects a single state, California's cap-and-trade program covers a population and economy roughly the same size as the multistate RGGI program and far more emissions due to its inclusion of the transportation sector.

Following the enactment of AB 32, CARB started a scoping process on options for meeting the law's goals, including a possible cap-and-trade program. This process included consultation with state environmental justice groups, which were opposed to emissions trading due to the risk of increasing concentrations of air pollutants associated with the burning of fossil fuels in poor and minority neighborhoods (Farrell and Hanemann 2009). Even as RGGI was completing its first auction, CARB (2008) released a final scoping plan that recommended a statewide cap-and-trade program for multiple sectors of the economy, including electricity generation and transportation, and multiple GHGs rather than just CO_2. This made the planned program one of the most ambitious in the world.

Two years later, CARB's Economic and Allocation Advisory Committee (2010, 3–4) recommended that the state auction most or all allowances under the program, and distribute the auction revenue directly to households along with using it to finance investments in emissions reductions "and other public expenditures." The committee was divided on whether allowance revenue should be returned to households in the form of income tax reductions or as dividend checks mailed to each household.

Prior to implementing the recommended cap-and-trade program, however, California had to withstand new political challenges reflecting the growing controversy over climate policy as well as cap and trade. Power generators and large energy consumers promoted a ballot measure to suspend implementation of AB 32 in 2011, arguing that the policy would be too damaging economically. Despite major funding from industry groups, the ballot measure was soundly defeated by a campaign stressing AB 32's ability to combine climate change mitigation with green economic development (Rabe 2015). In addition, environmental justice groups sued CARB in 2009 over concerns about the program concentrating pollutants in poor and minority communities, briefly delaying the program's implementation (London et al. 2013).

After these obstacles, CARB implemented the statewide cap-and-trade policy in 2012. In its final regulations, CARB (2013b, sec. 98570) set aside free allowances for a variety of groups including utilities that generate electricity and industries facing competition from companies in areas without emissions caps. The state then committed to auctioning the remaining allowances, dedicating that revenue to support the goals of AB 32 by funding a variety of programs designed to reduce the state's GHG emissions. In addition to these "state auctions," a 2012 ruling by the California Public Utilities Commission (CPUC) required investor-owned utilities to dedicate any value from their allowances to their ratepayers. This means that those utilities must also auction all the allowances they are given for free each year in what are referred to as "consignment" auctions, also conducted by the state. Including consignment auctions, more than 50 percent of all allowances in California in 2013 were sold at auction (Cliff 2014; see also Sekar, Munnings, and Burtraw 2014).

According to the CPUC (2012, 3), utilities must dedicate a large majority of the revenue from consignment auctions directly to ratepayers "on an equal per residential account basis delivered as a semi-annual, on-bill credit." The first of these dividends, amounting to approximately $35 per family, showed up on consumer bills in spring 2014 (Navarro-Treichler 2014). Some heralded the CPUC ruling as the first example of a true cap-and-dividend program with equal cash payments of auction revenue to all households (ibid.; Sandler 2012). California's direct rebate also represents an important alternative to the practice in the EU ETS and RGGI of spending these funds on primarily energy efficiency or other programs that mitigate climate change.

Interestingly, these direct consumer rebates have gotten relatively little political notice. More attention has been paid to the debate over how to spend the state auction revenue—a process that has generated enough political controversy to lead some to conclude that it threatens the program's long-term political viability (e.g., Rabe 2015). Under the CARB (2013a) regulations, the governor and state legislature must decide how to spend the state's share of the auction proceeds as part of the annual budget process. Because of AB 32's diverse statutory goals, political conflict has been fierce over a wide range of proposals for spending these state funds (Gallucci 2012a).

In response, the state legislature passed two laws in 2012 to better specify projects eligible for auction revenue. One bill (AB 1532) still permitted so many types of expenditures that it failed to narrow the field of potential funding projects much from the original AB 32 language (Rabe 2015). The other (SB 535) required that 25 percent of auction revenue be dedicated to projects benefiting "disadvantaged communities," but left it to state agencies to decide how to define which communities met that standard. Thus, neither bill did much to limit debate over how to spend auction revenue, as is attested by the first "cap and trade auction proceeds investment plan" issued by CARB in 2013. The document recommended spending most auction revenue on programs to promote "sustainable communities and transportation" programs, with fewer funds dedicated for energy efficiency programs, and a smaller amount proposed for agriculture and natural resource management programs (CARB 2013a). Language in the 2013 "investment plan" also rarely discusses protecting electricity ratepayers, even when supporting programs to improve energy efficiency. Instead, the focus is primarily on ways to use auction revenue to reduce emissions, which covers a wide range of programs (ibid., 19–20).

Also in contrast with the public benefit framing in RGGI, environmental advocates seem more intent on promoting state auction revenue mainly as a way to lower emissions rather than reduce the policy's impact on consumers. A report on potential uses for auction revenue by the Environmental Defense Fund (2012), for example, describes in detail how investments in commercial and industrial energy efficiency programs as well as new mass transit options could reduce GHG emissions while paying little attention to consumer energy bills. Concern over protecting average consumers from rate increases is not prominent in the same way it was in RGGI during conversations about investing auction revenue.

Political discourse has followed the same path. An updated scoping plan issued by CARB in 2014 stressed a broad range of investment options for auction revenue in order to reduce emissions in a variety of ways. This has led to the California legislature arguing annually with the governor's office over specific projects to fund, including a controversial new high-speed rail line connecting Los Angeles to San Francisco as well as increases in mass transit and affordable housing near transit hubs, among many other ideas. To date, auction money has been spread among all these priorities based on the results of annual budget negotiations, although a rough funding formula for future years started to take shape in 2014 (Mulkern 2014). Meanwhile, newspaper editorial boards (e.g., "Editorial: How California Should Budget" 2014) and other observers (Morain 2014; Mays 2014) worry that the diffusion of auction revenues to so many projects with potentially long-term or limited impacts on reducing climate emissions or protecting consumers from higher energy prices is a serious problem.

New political controversy also surrounded the program's expansion into transportation fuels in 2015. Citing estimates of higher gas prices for consumers due to the requirement for fuel wholesalers to purchase emissions allowances at auction, opponents from both political parties lobbied for either a delay in the inclusion of transportation fuels or an exemption of those fuels entirely from the cap-and-trade program (Lifsher 2014; White 2014). In this debate, arguments over consumer price impacts were front and center, with a Democratic member of the state assembly asserting: "The cap-and-trade system should not be used to raise billions of dollars in new state funds at the expense of consumers, who are struggling to get back on their feet after the recession" (quoted in Lifsher 2014). Environmental advocates tried to deflect these objections by claiming that estimates of gas price increases were overblown, and that consumers would still spend less money on gasoline overall due to programs encouraging more efficient cars and other investments under AB 32 (Baker-Branstetter 2014; Kaye 2014). Here are public benefit contentions that are more similar to those in RGGI regarding electricity price increases than seen in the earlier discussions of state auction revenue from the power sector.

With the addition of transportation fuels, California now auctions most of its allowances in the cap-and-trade program. As the revenue from those auctions grows, the state continues to struggle with competing perspectives on how to spend this money. In spite of some recommendations to limit auction revenue to a narrower range of uses directly linked to helping consumers, such as the 2010 report by CARB's advisory board, California looks more like the European Union in dedicating auction revenue to a wider range of public programs designed to mitigate climate change. At the same time, the California program is also providing the first direct cash rebate to consumers from auction revenue, albeit only for revenue from a minority of all allowances auctioned under the program. By combining these two revenue allocation approaches, California's program represents an intriguing hybrid of the narrower and broader understandings of the appropriate range of public benefits to be supplied by auction revenue— one that is distinct from both the (primarily narrower) vision of RGGI and the (primarily broader) vision of the EU ETS.

This unique combination makes California's cap-and-trade program the highest-profile conflict to date over the proper allocation of auction revenue. As the program continues to expand, consumer impacts will become more salient, raising pressure on cap-and-trade promoters to show how auction revenue will protect consumers directly from higher energy prices. How that debate plays out should serve as an important indicator of the potential future influence of the public benefit framing, including what uses of auction revenue will be most consistent with the new frame.

The Emerging Political Question: How to Allocate Auction Revenue

Controversy over the appropriate expenditure of auction revenue in the RGGI states, the EU ETS, and especially California indicates how the continued expansion of cap and trade with auction may depend on a key detail: the allocation of auction revenue. Indeed, the fortune of future climate policies may hinge at least as much on the answer to the question of who will bear the costs of the policy as it does on how strongly the public "believes" in climate change science.

The emerging conflict over acceptable distributions of auction revenue under the public benefit frame evokes a more long-standing dispute in US culture over the meaning of public ownership. RGGI represents an egalitarian vision of public ownership of the atmospheric commons that emphasizes the need to return auction revenues to nearly all citizens in some manner, in the form of widely used subsidies for energy efficiency improvements. Similarly, California's equal rebate to all electricity consumers is even more consistent with this egalitarian vision of public ownership, closely resembling Barnes's (2001) sky trust idea. These more egalitarian interpretations of public ownership as "owned by the people directly" are consistent with the public trust doctrine, a common law principle that limits the government's ability to privatize or restrict access to certain lands and waters that are owned in trust for its citizens (Sax 1970). The doctrine mandates that all citizens have equal rights to access and benefit from certain public resources (Wood 2014)—rights that government officials may not violate.

An alternative view conceptualizes public ownership as simply "owned by the government," thereby giving elected officials freedom to spend auction revenue on a wider variety of goals. This account, by contrast, provides government more discretion in deciding what uses of public resources are most appropriate than the narrower "trusteeship" model. Many programs to manage public lands in countries such as the United States, for instance, follow this more discretionary approach. Similarly, more flexible approaches to auction revenue such as those taken in the EU ETS and California are closer to this discretionary model of government ownership. In this respect, the conflict over auction revenue can be interpreted as yet another skirmish in a long-running battle over the exact meaning of public ownership.

Consistent with that view, some legal scholars are promoting the public trust doctrine as the foundation for a more effective global carbon emissions reduction scheme. Supporters of the idea of "atmospheric trust litigation" argue that the doctrine gives courts the power to mandate emissions reductions for national governments in order to protect the atmospheric commons for all citizens, including future generations (Wood 2012, 2014). Although the initial focus of this effort was on lawsuits to force nations to adopt binding GHG emissions reduction targets, the movement has evolved to incorporate Barnes's (2001) idea of seeking payments from polluters to fund a variety of projects to "maintain and improve the atmosphere for the benefit of all shareholders, present and future" (Costanza 2015).

Although this debate over the proper uses of auction revenue under the public benefit frame is likely to continue, it is noteworthy that RGGI's choice to more narrowly limit that revenue to programs benefiting the public directly has its roots in long-standing common law principles protecting the rights of *all* citizens. A discussion of prominent cap-and-trade failures in the next section will also suggest that the more strictly egalitarian approach seems to inspire greater public support than the more discretionary alternatives, consistent with the enduring influence of the public trust doctrine.

Failed Efforts to Promote Cap and Trade with Auction

Even as cap-and-trade programs became more ambitious in RGGI, the European Union, and California, other cap-and-trade efforts failed in the face of growing partisan opposition to climate policies after 2008. Many US states that had made initial commitments to regional cap-and-trade policies prior to 2008 subsequently abandoned them, and US federal efforts to create a national cap-and-trade program failed in 2008 and 2010. Internationally, Australia went through a fierce political battle over a proposed cap-and-trade program with auctioning starting in 2008, ending with the cancellation of the program in 2014. As noted at the beginning of the chapter, these failures have led some commentators to conclude that cap and trade is no longer a viable policy option to address climate change, calling the policy "a far more marginalized policy tool" (Rabe 2015), or even claiming that the policy is in "wide disrepute" or has "died" (Broder 2010).

This section reviews these recent challenges for cap-and-trade policies to address climate change, considering their implications for the book's argument that public benefit framing has created a new model for a politically successful climate change policy. It finds that the political failures of these policies can be attributed to growing partisan conflict and a global recession that posed challenges for all climate change policies as well as to deviations from the public benefit model that successfully overcame those challenges in RGGI, the European Union, and California. Although these failures illustrate the growing obstacles facing any new climate change policy, they also suggest that public benefit framing remains one of the best options for promoting such policies even in the face of increased opposition from partisan and economic interests.

State-Level Backlash: US Regional Cap-and-Trade Programs after RGGI

In 2007, California went beyond its own ambitious state program to organize a group of seven western states and several Canadian provinces to participate in a larger cap-and-trade program known as the Western Climate Initiative (WCI). Created by a movie star governor in charge of a state with the world's sixth-largest economy, the WCI looked like the next big step in regional cap and trade. Initial plans went beyond RGGI to limit six GHGs from a wider range of emissions sources, much as California did in its own state program (WCI 2008a, 15). Like RGGI, the WCI (ibid., 8, 33) aspired to auction all allowances in the long run. And like RGGI, arguments about the "fair treatment" of emitters and resource users, and the need to protect consumers through energy efficiency programs and consumer rebates, were prevalent in public comments on the program's 2008 draft regulations (WCI 2008b). Starting in 2010, however, a series of newly elected and more conservative Republican governors chose to withdraw from the program (Klinsky 2013). As of 2015, the only partners still working on the WCI with California were several Canadian provinces, including Quebec, which was planning to link its cap-and-trade program to California's.

The collapse of the other regional cap-and-trade program to emerge on RGGI's coattails, the Midwestern Greenhouse Gas Reduction Accord (MGGRA), was even more dramatic. Much like the WCI, the governors of six midwestern states agreed in 2007 on a regional cap-and-trade program to reduce their collective GHG emissions. Planning proceeded through a detailed report on potential policy designs (MGGRA 2009) that recommended auctioning nearly all allowances. By 2011, however, the election of a new group of more conservative governors in these states slowed momentum for the program, as did growing protest by state legislatures as they became more aware of the details. New economic challenges from the recession starting in 2008 also made any new climate policy more difficult in this coal-dependent region. The result was abandonment of the program by 2012 (Rabe 2015).

Thus, regional GHG cap-and-trade programs in the United States have declined from a peak of nearly twenty-five states in 2009 to just ten states in 2015 (RGGI plus California)—a process of "reverse diffusion" of the cap-and-trade idea among states (ibid.). Although there is no doubt that the failures of the WCI and the MGGRA are indicative of the new challenges facing climate policies starting in 2008, it is also important to contextualize

the scale of these defections. California's economy, for example, is sufficiently sizable that its cap-and-trade program is one of the largest in the
world even without the participation of other potential WCI partners. At
the same time, these failures are a useful reminder that no political strategy is guaranteed to succeed in promoting policy change in light of strong
opposition from vested interests or partisan mobilization. In some regions,
such as the coal-rich Midwest, resistance to any climate change policy from
vested and partisan interests has been too strong for these initial versions
of the public benefit model to succeed.

National Backlash: The Rise and Fall of a Federal Cap-and-Trade Program

Many RGGI supporters were open about their desire for the program to
serve as a model for a federal policy, and Congress actively considered a
number of GHG emissions trading bills during the RGGI design process and
while RGGI was being implemented. Although one of these bills eventually
passed the House in 2010, none were enacted into law. This section reviews
the failure of these federal cap-and-trade bills, and their implications for
cap and trade with auction and the public benefit model more generally.

Congress considered the Climate Security Act (CSA) in 2007–2008. The
law addressed a "basket" of multiple GHGs, attempting to reduce them by
a total of 19 percent from 2005 levels by 2020. The bill was broadly consistent with the RGGI model in proposing to auction a large portion of
allowances, giving away only about 15 percent of allowances to GHG emitters (Pew Center on Global Climate Change 2008). Unlike RGGI, the CSA
awarded allowances and allowance revenue to a wide range of constituencies, including programs for public lands management and deficit reduction (ibid.). Here the CSA deviated from successful programs that restricted
auction revenue to a narrower range of public benefits, as discussed above,
and was criticized on this basis. Unlike these successful programs, the CSA
failed to be enacted.

The election of President Barack Obama in November 2008 gave cap-
and-trade supporters new hope for a federal bill. With Obama taking office
in January 2009, and the more environmentally friendly Henry Waxman
assuming the chair of the House Energy and Commerce Committee, a coalition of national environmental groups and corporations known as the US
Climate Action Partnership (USCAP) and its congressional allies decided
the time was right to put another cap-and-trade proposal before Congress

(Rabe 2010). The USCAP proposal served as a blueprint for the leading climate change bill to emerge from the 111th Congress, the American Clean Energy and Security Act (ACES), also referred to as "Waxman-Markey" after its lead sponsors in the House (ibid.; Skocpol 2013). Like the CSA, ACES followed the public benefit model in some respects but not others. Although the bill proposed to auction a substantial number of allowances (Pew Center on Global Climate Change 2009), it failed to limit auction revenue to programs that directly benefited a broad swath of the public—a decision that again seems to have handicapped the bill's supporters and strengthened its opponents.

During the House hearings on ACES, numerous witnesses argued in favor of free allocation, describing auctions as equivalent to a "tax" that would increase energy costs, with fewer speaking in favor of auctioning (Rabe 2010). Perhaps as a result, the final version of the bill that went to the House floor proposed auctioning a smaller initial percentage of allowances. After a contentious floor debate, the bill managed to pass the House on a largely party-line vote, with most Republicans voting against the bill—an indicator of the growing partisanship of the climate change policy issue at this time (ibid.).

Members of Congress then went back to their districts for summer recess, where they faced a highly organized grassroots effort opposing the policy (Skocpol 2013). In the wake of these protests, cap-and-trade supporters struggled to craft a compromise bill that could gain sufficient Republican support to overcome a threatened filibuster in the Senate. Alternative bills more consistent with the RGGI public benefit frame had been introduced, including the CLEAR Act promoting 100 percent allowance auctions with revenue returned to directly to citizens as well as a last-minute bipartisan compromise advocating a rebranded "reduction and refund" approach designed to return more auction revenue directly to ratepayers (Samuelsohn 2010). But these alternatives either emerged too late in the process or lacked sufficient support from environmental advocates to overcome the partisan opposition to any federal climate policy, and the Senate failed to pass Waxman-Markey or these alternatives (Skocpol 2013, 54). Since this failure in 2010, Congress has declined to grapple seriously with climate change policy, as the Obama administration instead pursued an administrative strategy to reduce GHG emissions through new EPA regulations that will be discussed below.

Although there is disagreement on the exact reasons why ACES failed, at least one leading observer credits a highly motivated and well-organized national grassroots movement focused on the potential economic costs of the bill for the middle class. In an exhaustive analysis, Theda Skocpol (2013) describes how the USCAP group failed to build sufficient grassroots support needed to overcome the pressure by conservative activists on many members of Congress. In contrast to those promoting President Obama's successful health care reform bill, says Skocpol, environmentalists failed to build a national coalition around a compelling narrative of the tangible, specific benefits a climate bill might bring to average citizens. In language that echoes RGGI's public benefit framing, Skocpol condemns USCAP for failing to include and promote a simple, specific benefit in its proposal that citizens could easily rally around. Without such a benefit, environmentalists struggled to respond to the assertions of conservatives that the program hurt average citizens by raising energy prices in a time of great economic difficulty.

In sum, although a majority of the allowances in both Waxman-Markey and the CSA were dedicated to consumer price relief, many other allowances went to a wide range of industries as compensation for economic harms or toward a variety of other interests besides those of consumers. By failing to provide a tangible benefit for the general public in their cap-and-trade proposals, supporters could not respond effectively to criticism from opponents about the economic harm their policies would do to working Americans. In this respect, Skocpol's assessment of the failure of ACES is in conflict with claims that the piecemeal allocation of allowances to various interests helps to make such a policy politically viable (e.g., Pearlstein 2009). Instead, both bills appear to have failed in significant part due to their deviation from the public benefit model by treating auction income more like general tax revenue than a resource dedicated directly to benefit the many public owners of the atmospheric commons.

International Backlash: Political Turmoil over Cap and Trade in Australia

After ratifying the Kyoto Protocol in 2007, Australia's ruling Labor Party struggled to enact a GHG emissions trading scheme in the face of opposition from the Liberal Party, which was skeptical of the need for any climate change action and opposed to a price on carbon emissions. Although the

Labor Party eventually enacted the Carbon Pollution Reduction Scheme (CPRS) in 2011, the result was an unwieldy hybrid of a carbon tax and cap-and-trade system that imposed a relatively high, fixed price on CO_2 and other GHGs, only shifting to a "true" cap-and-trade program with a fixed emissions cap and floating allowance price in 2015. Despite Labor's efforts to promote the new policy as providing significant relief for consumers through tax reform and other credits to compensate for higher energy costs, the de facto carbon tax proved to be unpopular. Controversy over the policy led to multiple leadership changes in the ruling Labor government, and eventually the loss of a 2013 general election to a Liberal-led coalition promising to "ax the tax." In July 2014, the CPRS was repealed.

On the surface, the Australian policy looks much like the RGGI model, albeit with more money dedicated to direct payments for citizens than for energy efficiency subsidies. Dating back to its first proposals for a CPRS in 2008, the government regularly stressed its plan to auction a majority of allowances, using "every cent [the government] receives from the sale of pollution permits to help households and businesses adjust and move Australia to the low pollution economy of the future" (Australian Department of Climate Change 2008, xviii). This translated into an initial legislative proposal to dedicate approximately two-thirds of auction revenue to household assistance, including money to reduce income and fuel taxes. Under these initial designs for the CPRS, the remainder of the auction revenue was dedicated primarily to compensation for energy-intensive industries, with a small portion set aside for subsidies for energy efficiency programs and other technological innovations under a "climate change action fund" (ibid., xlii–xliv).

Australia's efforts to use auction revenue to protect consumers from higher energy prices make the CPRS failure a serious challenge to the argument that climate policies following the public benefit model should be more likely to succeed politically. In this respect, it is important to recognize Australia as a "hard case" for any climate policy action. It was one of the last nations to ratify the Kyoto Protocol, and features a large domestic coal industry, an expanding population, and energy-intensive industries that make it one of the largest GHG emitters in the world on a per capita basis. As noted earlier, even the most effective public benefit framing campaign may still fail when confronting strong opposition, and that opposition was and is formidable in Australia.

Australia's emissions trading proposals also deviated from the public benefit model in several key ways. First, there is little evidence of the public benefit frame in debates over the CPRS. Government proposals instead described auction revenue for households in terms of public assistance, much of it means-tested, to help Australians cope with higher energy prices (Australian Department of Climate Change 2008, xxxiii). By promoting revenue sharing with consumers using the language of assistance rather than entitlement, the government failed to fully utilize the power of normative frames portraying the atmosphere as a public good *already owned* by all citizens. Language in the actual CPRS bill also eschewed the public benefit frame, justifying auctions in terms of promoting "allocative efficiency" and "efficient price discovery," harking back to ineffective economic arguments for auctions under the old cap-and-trade model (Australian House of Representatives 2009, 102).

In addition, the mechanisms for distributing these funds differed from the public benefit model. The CPRS used auction revenue to lower income and fuel taxes, rather than funding energy efficiency programs or providing annual consumer rebates. Although reducing income taxes is a favored approach for using auction revenue among economists, it poses challenges for generating public support. Income tax reductions do not directly reduce consumer energy bills, unlike RGGI's programs subsidizing energy efficiency improvements for consumers or California's direct on-bill credits. This means consumers may still focus on higher energy bills from the emissions cap, and fail to connect them to their lower income tax levies. Australia's plan to use auction revenue to reduce per liter fuel taxes was also ridiculed as self-defeating because unlike a fixed dividend, it undermined the price incentive to reduce the use of carbon-based transportation fuels.

Perhaps most important, the CPRS differed from the new model of cap and trade: it was effectively a carbon tax, and a relatively high one at that. Several versions of the CPRS with an initial fixed allowance price of approximately AU$10 failed in 2009 and 2010, due to growing hostility toward any climate change policy among the (politically conservative) opposition Liberal coalition and misgivings about any policy designed to increase energy prices. In October 2009, for instance, a leadership challenge deposed Liberal Party leader Malcolm Turnbull, who had just agreed in principle to a compromise on the CPRS. The new Liberal leader, Tony Abbott, was more skeptical of the need for climate policy action, and condemned the CPRS as the Labor Party's "great big tax on everything" that would seriously harm Australian families and businesses alike (Abbott 2010). In the face of these hurdles, the CPRS could not get through Parliament, again consistent with the growing partisan opposition to climate change policies in many nations at this time.

Battered by the conflict, Labor leader Kevin Rudd announced in 2010 that the CPRS would be delayed until the end of 2012, and soon after lost his own leadership challenge. New leader Julia Gillard managed to avoid a defeat in the August 2010 elections in part by promising not to impose a carbon tax. Creating a coalition government with the Green Party (which greatly favored a price on carbon), Gillard remained prime minister on the condition that she would create a new committee to consider options for climate policy—the Multi-Party Climate Change Committee.

In July 2011, Labor issued a new framework for an emissions trading system titled "Securing a Clean Energy Future." Building on the deliberations of the Multi-Party Climate Change Committee, the document claimed to respond to criticisms of the original CPRS proposals, but still resembled the previous policy design in several fundamental ways. The new program again started with a fixed carbon tax and transitioned to a cap-and-trade program after a few years (which some considered a violation of Gillard's campaign promise). This decision to begin the program with a fixed tax would continue to have negative repercussions, as the tax failed to adjust downward in the face of weaker economic conditions the way an allowance price would under a cap-and-trade design.

In addition, the proposal continued to use public *assistance* framing to justify allocation of a majority of auction revenue to households. "A household assistance package will benefit millions," said the new proposal, "and will compensate those who need it most to cope with the cost of living impacts of a carbon price" (Australian Department of Climate Change and Energy Efficiency 2011, xiv). This framing of auction revenue as compensation for the *public* is the opposite of the language suggested by the public benefit frame used in RGGI, where *polluters* were compensated for transition costs, while the public received the well-deserved benefits from the private use of "their" resource. Finally, although the new plan claimed that a "large proportion" of industry and household assistance would go toward energy efficiency improvements, the plan's budget indicates that less than

5 percent of the auction revenue was actually dedicated to energy efficiency programs (ibid., 131–132). Thus, the Australian plan failed to provide direct bill assistance, funding for energy efficiency programs, or monies for significant other programs to reduce GHG emissions—an important distinction from the successful cap-and-invest programs in the United States and the European Union.

In November 2011, the Australian government finally passed the Clean Energy Act based on the principles in the new government plan. The new law created a CPRS with a three-year fixed price for allowances of AU$23 per ton (higher than earlier CPRS proposals, and much higher than EU ETS allowance prices at that time), scheduled to transition to a program with a fixed cap and floating allowance price in 2015. As with earlier versions of the CPRS bill, the legislation continued to describe the primary purpose of auctions to be allocative efficiency and efficient price discovery, mentioning the use of revenue for "other policy objectives" as a secondary motivation (Australian Senate 2011, 120).

When the new law took effect on July 1, 2012, the Gillard (2012) administration issued a press release promoting the implementation of the law's tax cuts to help taxpayers. Unfortunately for Gillard and her party, the new tax proved to be extremely unpopular during a time of poor economic performance in Australia. The fixed allowance price, which was initially proposed as a way to limit the program's economic costs, was now far higher than the price of carbon in other developed countries. Ironically, even as Europe worried about allowance prices that were too low, Australia's comparatively high allowance prices created even greater controversy.

As a result, pressure built against the Australian ETS. Within a year of the tax being introduced, Rudd (2013) reassumed the prime minister position, and announced plans to shift to the flexible price version of the ETS a year early to provide price relief to consumers and industry. Despite these efforts, Labor lost the September 2013 elections to Abbott's Liberal coalition, based substantially on opposition to the ETS program. On winning, Abbott's party immediately vowed to repeal the ETS, while Labor continued to argue that its plan to shift to the floating price structure under a "true" cap-and-trade model was the preferable option. Abbott eventually was able to repeal the emissions trading program in July 2014, but plans for a long-term replacement policy remain controversial (Taylor 2014) with cap and trade continuing to be part of the discussion (Arup 2015; Lim 2014).

In sum, supporters of cap and trade with auction in Australia made several decisions that deviated from the successful public benefit model found in RGGI, the European Union, and California, including starting with a much higher fixed price for allowances, failing to dedicate auction revenue to household energy efficiency programs or other broadly egalitarian benefits, and failing to frame the policy in terms of polluter pays and egalitarian norms of public ownership. In this respect, Australia's experience suggests not only how difficult promoting any significant climate change policy has become in some parts of the world since 2008 but also how justifying even cap and trade with auction in ways that deviate too far from the public benefit model appear to be less politically effective.

Australia also is a reminder of the limits of normative reframing as a strategy for political change. As noted in chapter 2, norms vary by society, and it may be the case that the relevant egalitarian and polluter pays norms are less widely supported in Australia than in the United States. Moreover, although normative reframing can improve the chances for policy change, it does not make that change inevitable. It could be that in Australia, organized opposition to serious climate policy action is so strong that a normative reframing strategy based on public benefit framing is insufficient to create stable policy change on this issue at this time.

Public Benefit Framing for Other Climate and Environmental Policies

Public benefit framing has the potential to promote other policies besides cap and trade with auction. Logically, the public benefit frame relies on the combination of two fundamental norms: that polluters should bear the costs of their use of a public resource, and that the public should benefit broadly from that use. In the case of climate change, the resource in question is the atmospheric commons: the ability of the atmosphere (and linked terrestrial systems) to absorb a limited amount of GHGs. In the case of carbon pricing policies, the potential public benefits include *consumer benefits* provided by direct rebates or subsidies for programs to reduce consumer energy demand from high-carbon sources (as emphasized in RGGI) as well as what one might call *climate protection benefits* from investments in infrastructure and research to reduce future climate change impacts (as stressed in the European Union). More recently, a third potential public benefit from climate mitigation policies has been promoted in the form of *public health benefits* from the reduction of more traditional air pollutants associated with GHG emissions. This third variation of the public benefit frame has proved especially important in US federal climate policy since the failure of the Waxman-Markey bill, as will be discussed below.

A wider range of public benefits implies a wider range of policies that could be justified with this particular normative reframing. Alternative carbon-pricing policies, such as a revenue-neutral carbon tax, are easily associated with public benefit framing in terms of consumer benefits or climate protection benefits, albeit with potential pitfalls as seen in the Australian case. Even policies reducing GHG and associated copollutants without a pricing mechanism might also fit under a public benefit frame, especially one that stressed public health benefits. Finally, other environmental policies could be framed as generating crucial public benefits from the private use of a scarce public resource. This section will look at examples of all three types of additional applications of the public benefit frame.

Public Benefit Framing for Other Carbon-Pricing Policies: British Columbia's Carbon Tax

As noted, carbon taxes that redistribute revenue to citizens are close to the RGGI model of cap and trade with auction, making them an obvious candidate for public benefit framing. Carbon taxes have their own political challenges, however. By more directly raising the costs of vital services such as heating, cooling, and transportation for all consumers, they invite stronger public criticism than cap-and-trade policies, and are relatively uncommon globally. Successful proposals to impose carbon taxes have typically required a dedicated political champion willing to pursue the idea in the face of strong misgivings by his or her political party and inconsistent public support (Harrison 2010, 2012). In addition, early carbon taxes in Denmark, Finland, and Germany all included substantial exemptions for politically influential industries (Harrison 2010), consistent with interest group models of politics that explained the failure of cap and trade with auction for decades.

Nevertheless, there is evidence that public benefit framing can increase the political viability of a carbon tax. Research has suggested that the public is more supportive of environmental taxes that earmark revenue directly toward addressing the environmental problem in question (ibid.; Lachapelle, Borick, and Rabe 2012)—a policy design consistent with the successful public benefit model in RGGI, the EU ETS, and California. A more detailed look at British Columbia's carbon tax, arguably the most successful carbon taxation policy to date, confirms the applicability of the public benefit frame to this additional policy design.

After more than a decade of limited interest, the Canadian public made environmental concerns a higher priority in 2006. The transformation generated a number of proposals to address a range of environmental issues including climate change—a topic the Canadian government had largely ignored despite ratifying the Kyoto Protocol (Harrison 2012). Prominent among these efforts was a surprising proposal by BC premier Gordon Campbell to introduce a revenue-neutral carbon tax. Although Campbell's motivations have been the subject of much speculation, his discussions with California governor Arnold Schwarzenegger about AB 32 and the WCI appear to have influenced his thinking (ibid.).

In February 2007, Campbell formally proposed the idea of a 33 percent reduction in the province's GHG emissions by 2020. He also introduced several potential policies to achieve this goal, and asked individual ministries to submit additional ideas. That summer, the Ministry of Finance formally suggested the idea of a carbon tax, and grassroots support for a revenue-neutral tax quickly grew. Even the business community did not object to the idea as long as the tax did not try to redistribute income or generate income for new projects (ibid.).

Less than a year later, on July 1, 2008, the new carbon tax was implemented at a rate of C$10 per tonne of CO_2 equivalent emissions for carbon-based fuels (British Columbia Ministry of Finance 2013, 63–66). Scheduled increases of C$5 per tonne occurred annually, until the tax reached its final proposed level of C$30 per tonne on July 1, 2012 (ibid., 63–66). Revenue is returned entirely to taxpayers in the form of income and corporate tax reductions, with the original distribution being two-thirds to individuals, and one-third to corporations. Furthermore, the government inaugurated the tax with a onetime "climate action dividend" payable to all BC residents in 2008 (Rabe and Borick 2012).

Although the public's initial reaction to the tax proposal was positive, an unrelated increase in gasoline prices prior to the tax's implementation in 2008 reduced that support (Harrison 2012). The tax also faced yet another ax-the-tax campaign, ironically launched by the traditionally more environmentally friendly New Democratic Party (Shaw 2011). This campaign

against the alleged unfairness of the carbon tax made some political head-
way, with polls indicating that many residents did not believe the tax was
truly revenue neutral (Harrison 2012). At the same time, the Liberal Party
successfully retained control of the provincial parliament in 2009 without
backing off the carbon tax, and by 2014 both parties were on the record
as supporting the policy. A public review of the program in 2012 included
more than two thousand public comments indicating continued support
for both the carbon tax overall and its revenue neutrality, leading the government to maintain the tax without substantial revision (British Columbia
Ministry of Finance 2013, 63–66). This is consistent with public opinion
polls suggesting the policy has gained support over time since its adoption
(Rabe and Borick 2012).

Thus, the BC experience confirms several important points about the
politics of carbon pricing. First, it appears to be true that placing a direct
cost on carbon through a tax remains more politically challenging than
imposing that cost more indirectly via a cap-and-trade program, much as
was the case in Australia. The case also confirms the political significance of
dedicating revenue from a carbon-pricing mechanism directly and clearly
to public benefits in order to maintain support for the policy. Third, the
initial controversy and confusion over whether the tax was truly revenue
neutral speaks to the challenge of getting consumers to connect income tax
reductions or other more indirect forms of revenue recycling with highly
salient increases in energy costs, as also happened in Australia. The efforts
of the BC government to limit any use of carbon-pricing revenue entirely to
protecting consumers nonetheless seems to have been critical to the long-
term viability of the policy (Rabe and Borick 2012), consistent with the success of the public benefit model in other settings. When the costs imposed
on consumers by the policy are even more obvious, as in the case of a carbon tax, the public benefit frame along with its implied arrangements for
allocating revenue appears to be even more crucial.

Public Benefit Framing Beyond Carbon Pricing: US EPA's 2015 Clean Power Plan

A critical test of the ability of public benefit framing to promote different
climate policies in different political contexts arrived in the form of EPA
regulations for GHG emissions issued in August 2015. Under its "Clean
Power Plan" (CPP), the US EPA (2015d) required states to submit plans to

reduce their GHG emissions from power plants by an average of 32 percent from 2005 emissions by 2030 or accept direct federal regulations on the state's power generating facilities. State GHG targets were set by EPA analysis of each state's ability to reduce its emissions according to three "building blocks," including potential efficiency gains at coal-fired power plants, the potential to shift more electricity production to natural gas facilities, and the potential to increase the share of renewable energy in the state (ibid., 64717). New allowable emissions rates were benchmarked according to two types of facilities instead of being adjusted from historical emissions levels, increasing the required reductions for higher-emitting power plants (Ramseur and McCarthy 2015). State targets range from a 7 to 47 percent decrease in emissions from 2012 levels by 2030, with the highest reductions in states most dependent on coal-fired production of electricity ("E&E's Power Plan Hub" 2015).

The EPA rules give states broad latitude for deciding how to meet the new federal GHG reduction targets (US EPA 2015a). States can simply enforce the new federal emissions rates on their generating units, consistent with earlier command-and-control regulatory approaches. Alternatively, they can take a "state's measures" approach that allows them to adopt a variety of policies to meet the same overall emissions goals. Acceptable state policies include creating an intrastate or interstate cap-and-trade program, with or without an auction, and with or without investment in energy efficiency programs, as well as other policies such as "renewable energy standards" requiring a minimum level of energy production from renewable sources (US EPA 2015a). As such, the range of policies being promoted by the EPA and CPP advocates goes beyond carbon pricing.

Despite these efforts to increase state flexibility, a number of states have strongly opposed the CPP. A group of these states sued unsuccessfully to oppose the CPP in 2015, and more lawsuits have followed, including one that resulted in a temporary stay of the program in 2016. Meanwhile, some Republican governors have threatened to refuse to implement the proposed CPP rule, and traditional fossil fuel interest groups continue to complain about what they term the EPA's "war on coal" (Groppe 2015; Holden, Kuckro, and Behr 2015).

As in RGGI and California, a primary argument against the plan is its potential effect on energy prices (Groppe 2015). Even prominent climate science skeptic Senator James Inhofe chose to stress the plan's negative impact on consumers, calling the plan "the most regressive of all taxes ever passed" and claiming that it would cost "about $3,000 for every taxpaying family in America" (quoted in Corombos 2015). Other reactions to the CPP by conservative think tanks echoed this message, comparing the plan to the failed Waxman-Markey cap-and-trade bill, and condemning its impact on middle-class Americans (e.g., Loris 2015). In this respect, state responses to the CPP are an important potential test of the ability of public benefit framing to increase support for new climate policies that are being condemned by partisan and energy interests for harming consumers by raising energy prices.

Based on the analysis provided in this chapter, there is good reason to think that public benefit framing is still the best option for promoting a successful state policy response to the EPA mandate (or similar future efforts should the CPP itself be blocked by the courts), even in the face of this tougher political opposition. Perhaps accordingly, the EPA and environmental advocates have drawn on the public benefit model to promote the CPP, both in the recommended policy design for compliance and the framing used to promote state compliance with the federal rule.

Cap and trade, including interstate cap-and-trade programs resembling RGGI, was a prominent part of the draft rules for the CPP in 2014, and the final rules included several changes that made interstate emissions trading programs even easier for states to adopt (Upton 2015). This emphasis on making it as easy as possible for states to be "trading ready" in their plans (e.g., Litz and Macedonia 2015) has also been accompanied by public discussion of states joining existing cap-and-trade schemes such as RGGI's or California's program (Kahn 2015; Holden 2014).

The EPA's proposed model rule for complying with the CPP relies on a cap-and-trade approach. Released simultaneously with the CPP, the model rule offers two options to states: either a "rate-based" plan that requires electricity-generating units to emit below a fixed rate per unit of energy produced, or a "mass-based" approach that works like a traditional cap-and-trade program, requiring these electricity-generating units to submit emissions allowances for every unit of CO_2 they emit (US EPA 2015c). Because the rate-based option allows firms to trade and use "emissions reduction credits" to meet their emission rate goals, it also incorporates market principles that make it look like a cap-and-trade program. The EPA (ibid., 64970) also announced its intention to select either a rate- or mass-based approach only for the final model rule in 2016, and expressed its preference for a mass-based rule in some detail in the draft regulations. In this regard, the model rule is widely viewed as creating an interstate cap-and-trade program, thereby establishing a RGGI-like program as the path of least resistance for states not wanting to enact their own plans.

The proposed federal rule does not follow the new cap-and-trade model entirely. It awards allowances to generating units for free, for instance, based on their historic levels of energy production in the state (ibid., 65016). This "output-based" allocation is not a reversion to grandfathering allowances according to prior emissions levels, but it is a free distribution of allowances to emitters, which is contrary to the policy design associated with the public benefit model. Although this is an important potential exception to the trend toward greater auctioning, it is driven in part by the EPA's (ibid., 65018) concern that revenue from any federal auction of allowances would legally be required to go to the federal treasury rather than to the states implementing the program.

Additional discussions of allocation in the model rule are more consistent with the public benefit approach. The rule encourages states to consider their own allocation plans even if they adopt the rest of the federal rule without revision, and promotes auctions as an allocation option, mentioning RGGI by name as an example of using auction revenue to fund energy efficiency programs "to help reduce electricity rate impacts" and program costs (ibid., 65018). In addition, the model rule sets aside some allowances for new sources of renewable energy in order to encourage the development of these low- or zero-carbon power sources (ibid.). In short, even though the CPP and associated model rule does not mandate auctions, in other respects it strongly resembles the new model of cap and trade by encouraging the use of allowances to lower costs for consumers and to subsidize the development of sources of low-carbon energy.

EPA officials have also emphasized public benefit framing in making the case for the CPP. Some of this framing relies on the consumer protection benefits that were central in RGGI. As EPA administrator Gina McCarthy (formerly a leading player in RGGI) has argued publicly, "Critics claim that your energy bills will skyrocket [under the CPP]. They're wrong. Should I say that again? They're wrong" (quoted in Cushman 2014). Here McCarthy and other EPA officials (e.g., US EPA 2015b) have promoted the ability of potential investments in energy efficiency and similar consumer programs to actually reduce consumer electricity consumption and energy bills—arguments that come directly from the RGGI playbook. Environmental advocates working on implementation of the CPP and related energy and climate issues in Republican-controlled states such as Indiana also cite impacts on consumer electricity rates as relatively more important to conservative lawmakers than arguments about economic development or public health threats (e.g., Kharbanda 2015). Here again, one can see that the consumer protection ideas emphasized by the public benefit frame are more likely to speak to concerns of politically conservative individuals, as was first suggested in chapter 2.

At the same time, defenders of the CPP have relied even more strongly on a new type of public benefit framing underscoring the public health benefits for all citizens due to cleaner air from the reduction of the copollutants associated with GHGs. Consider this passage from the US EPA (2015a) overview of the CPP:

> The Clean Power Plan cuts significant amounts of power plant carbon pollution and the pollutants that cause the soot and smog that harm health, while advancing clean energy innovation, development and deployment, and laying the foundation for the long-term strategy needed to tackle the threat of climate change. By providing states and utilities ample flexibility and the time needed to achieve these pollution cuts, the Clean Power Plan offers the power sector the ability to optimize pollution reductions while maintaining a reliable and affordable supply of electricity for ratepayers and businesses.

Although this summary mentions consumer protections such as keeping electricity affordable for ratepayers and the climate benefits from investing in low-carbon options, the featured benefit is the reduction of "the pollutants that cause the soot and smog that harm health." Other EPA documents (US EPA 2015b) highlight the specific projected public health benefits of the CPP, including reductions in premature deaths, asthma attacks, hospital admissions, and missed workdays caused by air pollution. Public statements by senior EPA officials have also stressed these public health benefits from reduced emissions of "carbon pollution" and associated pollutants from burning fossil fuels, often above and beyond other risks from climate change. In a June 2014 speech on the CPP, EPA administrator McCarthy (2014) delivered an anecdote about a ten-year-old boy with asthma, ending with this assertion: "Carbon pollution from power plants comes packaged with other dangerous pollutants like particulate matter, nitrogen oxides, and sulfur dioxide, putting our families at even more risk." Even President Obama (2015) stressed the public health benefits of the CPP's reduction of copollutants in announcing the plan on August 3, 2015.

This emphasis on more immediate public health improvements from reducing emissions associated with the burning of fossil fuels is a relatively new way of justifying climate change policy action. Although the idea of improving local air quality by reducing emission of the copollutants associated with GHGs is not a new one, the argument had limited political salience until recent discussions over the CPP. Rather than focusing on the economic gains for consumers from charging polluters for their use of the atmosphere, this approach emphasizes the widely distributed health gains for most citizens from cleaner air. Such an idea seems to fit well with the egalitarian norm that supports public benefit framing, and it addresses environmental justice concerns about unequal concentrations of traditional air pollutants in low-income and minority communities—a frequent criticism of cap-and-trade policies. At the same time, the connection to the polluter pays norm is less obvious. For this reason, it will be interesting to see how this particular variant of public benefit framing influences policy debates compared to arguments and policy designs more focused on charging polluters for their use of the atmospheric commons, and using that revenue to lower consumer energy bills or invest in programs to lower carbon emissions more broadly.

A focus on public health benefits also widens the range of policies that might be seen as consistent with the new public benefit framing. In the case of the CPP, a variety of policies could reduce GHG and associated pollutants in a manner consistent with these health benefits, including cap and trade without auction, or even so-called renewable portfolio standards that require electricity generators to increase the portion of energy produced from low- or zero-carbon-emitting sources.

Thus, the CPP represents an important new effort to promote new climate policies, including but not limited to cap and trade with auction, on the basis of public benefit framing revolving around *both* consumer protections and the newer idea (for climate policies) of public health benefits. If the CPP survives its various legal challenges, many observers believe the ensuing state policies will represent a major expansion of cap-and-trade limits on GHG emissions. Partisan and interest group objections remain strong, however, and may prevent the auctioning of allowances in some states even under a cap-and-trade plan. How states respond to this challenge, and the ability of different issue frames to influence those state policy choices, will be a good test of the limits of public benefit framing's ability to promote different climate change policies even in the face of strong opposition.

Public Benefit Framing for Other Environmental Policies

Public benefit framing could apply to many other private uses of public resources besides emissions of GHGs. Policies for air pollutants other than GHGs, for instance, might be more politically acceptable if they required emitters to pay for the right to put these pollutants into the atmosphere, and dedicate that revenue to programs generating broad public benefits offsetting any potential increase in energy or other costs for the public. Although such policies have to be careful to avoid concentrating harmful air pollutants in high-pollution neighborhoods, the public benefit framing makes this obligation more explicit by requiring the program to benefit *all* citizens, including those located in the most polluted areas. Such programs could also dedicate some revenue to investments in technology to reduce these locally harmful emissions or develop alternative sources of energy that emit fewer air pollutants of all types to address these environmental justice concerns, as has occurred to a limited degree in California under AB 32.

Efforts to promote cap and trade or similar market-based programs to reduce water pollution are another promising venue for public benefit framing. This would involve portraying waterways as public resources that private actors must pay to use for their pollution as well as dedicating that revenue to specific and tangible public benefits. Public benefits in this case could include bill rebates or subsidies for water efficiency to reduce household water consumption. Or pollution charges could fund public investment in costly renovations to municipal sewer systems that discharge untreated sewage into waterways during large storm events or new technologies for reducing other sources of nonpoint-source water pollution. By designing and promoting these programs based on creating such broad public benefits, while making polluters pay for their use of this resource, such programs might gain greater political viability.

The public benefit frame also shows promise for policies related to the allocation of extractive rights to certain public resources, such as fisheries, timber, or rangeland. In many marine fisheries, for example, grandfathering existing use rights remains a common practice in the creation of new market-based policies using "individual transferable quotas" or "individual fisheries quotas" to limit access (Hannesson 2004; Macinko and Bromley 2002). Such policies could instead auction these new private rights, dedicating that revenue to a variety of important public benefits, as opposed to giving them away in a manner that often favors boat owners over others who have worked in the community on fishing for decades (Macinko 2010). Similar arguments could be made about long-standing US federal policies providing public range forage, timber, and water at below-market costs to private actors.

Finally, the public benefit frame could be extended to policies relating to the extraction of valuable oil and minerals. As discussed in chapter 3, the federal government and some states have long imposed severance taxes on the extraction of oil and certain types of minerals from public lands. More recently, other states have adopted similar severance taxes, especially on shale gas deposits. Except for Alaska, however, the federal government and most of these states treat the revenue from these taxes as general tax income rather than dedicating it to specific environmental or public benefit uses (Rabe 2014). How much governments will continue to charge firms for the private extraction of these public resources, and what they will do with those funds, remains an intriguing future question, even as such policies remain strong candidates for public benefit framing strategies in their future promotion and design.

Using Normative Reframing in Other Policy Contexts

Many policies are justified by different normative frames, and therefore potentially vulnerable to sudden policy change through a normative reframing strategy. This raises the question, How can advocates and researchers identify other policies vulnerable to policy change through a normative reframing strategy, beyond the examples related to the public benefit frame discussed so far?

As was explored in detail in chapter 2, the overall influence of a normative frame is a product of the force of the norm being cited by the frame—that is, the norm's degree of influence over behaviors and attitudes—and the norm's perceived fit with the issue. More forceful norms exercise greater influence over individuals' behaviors and attitudes, permitting fewer violations and generating greater sanctions by others for those violations. In this sense, frames portraying a policy as supported by a more forceful norm should increase public support for that policy, making the policy relatively more stable than one framed in terms of a weaker norm. Frames also vary in their perceived fit with an issue—that is, their ability to convince individuals that the frame is appropriately applied to a particular issue. In the case of normative framing, one can think of variations in a frame's influence as a product of the strength of the norm being used and its perceived fit with the policy position being justified. If one can document both the general strength of the norm being used to support a given policy and the norm's perceived fit with that particular issue, it should be possible to identify other policies that lack a strong normative foundation for support and hence are more vulnerable to sudden change.

The general process of identifying policies with relatively weak normative foundations is straightforward and relies on standard methods of social science research. As introduced in chapter 2, it starts by identifying the norm or norms that are dominant in the public framing of the issue. Here, interviews with key actors in the policy debate combined with standard content analysis methods of public and private discourse will indicate the relative prevalence of various frames among those debating the issue. By determining which norms are most frequently appealed to, either explicitly or implicitly, by those defending the policy status quo, the researcher can identify the key norms supporting the policy. Through process-tracing techniques, including in-depth interviews with those involved in the policy process and careful consideration of counterfactuals (Collier 2011; Collier, Brady, and Seawright 2010), one can better evaluate the apparent influence of those particular norms on preserving the status quo.

The details of this process will vary, of course. In many cases, it will be important to look at the prevalence of different norms in the policy debate over time. This analysis would start with documenting the normative frames present at the initial public debate when the policy was first adopted, when those frames are more likely to be prominent. Here the researcher can determine which norms were cited by different supporters and opponents of policy change as well as which norms were most influential over the final decisions made. In addition, content analysis might continue to look at more recent discussions by elites and interested stakeholders, and the frames they now use to describe the policy compared to those present at the policy's adoption.

Ideally, such work would rely on both public and private discourse, including public conversations about the issue in the media, public hearings, and floor debates as well as private meeting records or interviews about the issue with key participants. Coding could include either a qualitative analysis of the presence of various frames citing different norms, a quantitative process looking for key words or phrases as evidence of those normative frames in the record, or both. Previous research has frequently documented the relative prevalence of various frames for different policies, including both for policies actively under debate and those that are more stable (e.g., Klüver and Mahoney 2015; Delshad and Raymond 2013; Borah 2011; Clawson and Trice 2000). The point is simply to identify the norm or norms most widely relied on by those advocating the current policy over time in both public and private discourse.

This work would also include standard process-tracing techniques to assess which norms appear to have been most politically influential in supporting the status quo. In some cases, the importance of a given norm may be indicated in interviews with those involved in the policy debate. In other instances, a particular norm's policy significance can be documented by "the dog that didn't bark" (Collier 2011)—that is, the lack of serious efforts to challenge a specific normative argument even by those who would benefit greatly from a different policy approach (e.g., Raymond 2003, chap. 3). Here counterfactual reasoning is useful: outcomes that are consistent with a particular norm in the public record but otherwise hard to explain suggest the greater importance of that norm in shaping the policy status quo.

Having identified the dominant normative frame(s) supporting a given policy, the next task is to document the relative force and fit of the norm or norms cited by the frame(s). Again, this work is already being done using a variety of techniques. For example, researchers use surveys of representative samples of the public to measure the force of different norms. Work of this nature may simply ask respondents about their support for different normative principles (e.g., Craemer 2009; Raymond and Schneider 2014), willingness to impose sanctions for different norm violations (Traxler and Winter 2012), or perceptions more generally of the relative strength of different norms in society (Crandall, Eshleman, and O'Brien 2002). Research can also document the relative force of different norms through efforts to associate belief in those norms with behavioral intentions or actual behaviors. Some of this work asks subjects to indicate how acceptable different behaviors are, such as discrimination in employment or housing (ibid.). Other work focuses more on observations of behavior, manipulating the decision context to increase or decrease the cost of violating a particular norm in order to identify how strongly that norm affects behavior. In this context, a more forceful norm generates behavior that is relatively more costly for experimental participants, such as the honest self-reporting of pollution (Raymond and Cason 2011) or failure to "defect" in certain cooperative games (Ostrom 1998). Experiments also can document a norm's relative force by how frequently other participants are willing to punish norm violators, even where those punishments are costly to the norm enforcer (ibid.; Fehr and Fischbacher 2004).

Alternatively, research can document the relative force of norms through observations of behavior in the field (e.g., Hechter and Opp 2001). Here again, evidence of a norm's relative force can be determined by observations of costly behaviors consistent with the norm such as failure to properly insure for liability for a livestock vehicle collision based on the (legally incorrect) norm that the "driver buys the cow" in many of these situations (Ellickson 1991). Other field studies document patterns of costly behavior in a variety of social contexts to indicate the existence of more or less forceful norms (Henrich et al. 2005). Finally, it is also possible to evaluate normative force through observations of costly punishments in the field for norm violations, ranging from low-cost sanctions like negative gossip to higher-cost punishments for more serious norm violations such as public shaming, physical violence, property destruction, or even exclusion from the community (Ostrom 1998; Elster 1989). Historical research can also provide insights into a norm's strength in a society over time, including evidence of changes in a norm's social influence.

Because norms will often conflict when applied to the same situation, it is also important to measure a norm's perceived fit with an issue, or how applicable individuals think one norm is to a given situation compared to an alternative. To answer this question, researchers use survey or interview techniques to elicit perceptions of a norm's relevance to a specific issue. Andrea Olive (2014), for instance, used interviewing to determine how well landowners felt private property and environmental stewardship norms fit with different policy requirements for protecting endangered species on their property, while Jennifer Hochschild (1981) used interviews to better understand how Americans applied different distributional norms to social policy dilemmas. Other work uses surveys or focus groups to evaluate how applicable individuals believe different norms are to an issue (Delshad et al. 2010), or how important different norms or values are to shaping a particular attitude held by the individual (e.g., Brewer 2003). One can also imagine observational research identifying the boundary conditions beyond which a particular norm appears not to apply (for example, for members of different groups, or for some decision contexts but not others based on variations in circumstance).

Researchers also assess a norm's perceived fit with particular issues experimentally by looking for variations in the application of a norm across different experimental conditions. If individuals tend to distribute some resources equally but not others, for instance, this suggests differences in the perceived fit of an egalitarian norm with different distributional tasks: distributing slices of birthday cake, say, versus income from the sale of a cake created through unequal contributions of effort (see generally, Stone 2002). Good illustrations of this type of work can be found in many simple distributional experiments already discussed, where adding conditions suggesting that a player has previously "earned" the sum of money that is then required to be divided between the two players leads to less generous offers, pointing to the greater applicability of a Lockean norm in this context (e.g., Franco-Watkins, Edwards, and Acuff 2013)

Having determined the norm or norms most responsible for supporting the current policy arrangements, and the degree of public support for that norm's application to this particular issue, the researcher is then able to determine if the policy features a relatively strong or weak normative foundation. If that foundation is relatively weak, either because the norm supporting the policy is less powerful or because it is not seen by many as "fitting" the policy in a convincing manner, then the policy is more vulnerable to sudden change based on a normative reframing strategy.

As an example, one could imagine this pattern of research regarding policies related to a controversial issue such as same-sex marriage. Initially, researchers would look at the rhetoric used to justify laws limiting marriage to one man and one woman, including any particular norms being cited about the appropriateness of consensual homosexual relationships as well as the social significance of marriage. Researchers might then use standard public opinion methods to measure public support for these norms, and the willingness to violate them or punish those who do. This work might also attend to changes in the perceived force of the norms over time, if any. In addition, they might use observational research to determine the openness of homosexual relationships in a society as well as public condemnations of such relationships. Similar observational and historical work might document changing patterns of marriage reflecting different norms about appropriate forms of child-rearing and family relationships.

Researchers also might investigate whether the public believes that norms about the morality of homosexual relationships are most relevant to, or fit with, the issue of gay marriage as strongly as alternative norms emphasizing the need to respect individual freedoms or favoring stable monogamous relationships. Research indicating either a weak level of support for the norms supporting current policies, or a weak perceived fit between those norms and the policies prohibiting marriages between same-sex couples, would indicate that this policy is ripe for change.

Events since 2010 suggest that normative reframing is a good candidate for explaining what has happened with regard to changing policies about gay marriage. On the one hand, evidence suggests that support for the norm against consensual homosexual relationships is weakening, especially among younger individuals (Baunach 2011). Moreover, the public seems less inclined to apply these particular norms about appropriate sexual behavior to the issue of permitting civil unions or marriages between two committed adults of any sex, opting instead for an "equality/tolerance" frame in many cases (ibid.; Mucciaroni 2011). There thus appears to be an ongoing shift in support for the moral norm underwriting the old policy status quo (that homosexual relationships are inappropriate) and the perceived relevance of a moral norm about sexual orientation to such policies. In this case, a weakening of both the *force* of the norm supporting the old policy status quo as well as its perceived relevance or *fit* with the issue of state-sanctioned marriage seems to have been central to the trend of new state policies allowing for same-sex marriages (prior to the US Supreme Court decision in 2015 requiring all states to recognize such marriages). This is in contrast to the RGGI case, where the primary factors shaping policy change was not a weakening of support for the Lockean norm per se but rather a sense that the norm failed to fit the issue of distributing private rights to emit pollution. In this sense, a policy's normative foundation can be weak either based on changes in the apparent force of the norm support-ing the policy or its perceived fit with the policy question, or both.

In sum, many policies rest on important normative foundations, and some of those foundations are measurably weaker than others. The strength of these foundations can be measured using existing methods to determine how significant a particular norm is to a given issue, and how forceful that norm is in society and how well it is perceived to fit with the issue. By studying these normative foundations, social scientists should be able to assess the apparent vulnerability of a wide range of policies to future change through normative reframing, in addition to explaining pre-vious policy changes. Indeed, this book makes the argument that the new carbon-pricing policies dedicating revenue to broad public benefits should be more politically stable in many contexts than other approaches to the problem of climate change, based on their stronger normative foundation. By bringing a new, testable set of hypotheses about what policies are more or less likely to experience sudden policy change in the future, the norma-tive reframing perspective offers a useful step forward in theorizing policy change across a wide range of issues beyond the climate policy examples discussed in this volume.

6   Conclusion

In 2003, a group of northeastern states seeking to reduce their GHG emissions created the idea of RGGI. Five years later, those states had developed the first cap-and-trade program to address GHG emissions in the United States, and the first policy in the world requiring polluters to buy virtually all their required emissions allowances at auction—a result few would have predicted. In this sense, the creation of RGGI was indeed a revolution within the world of environmental policy. After decades of assuming that emitters were entitled to use the atmosphere at no cost, in a few short years a group of policy advocates transformed the debate.

Under RGGI, allowances were reconceptualized from private entitlements for existing emitters into shared assets to be distributed for the public's benefit. In "reclaiming the atmospheric commons" in this manner, auction supporters drew on the power of norms to overturn a long-standing practice of giving away valuable rights to use a public resource to economically powerful interests. In particular, they increased the influence of the polluter pays norm by pairing it with an egalitarian norm to govern the appropriate distribution of those payments by polluters. By creating a new public benefit frame combining these two norms, environmental advocates better justified auctioning emissions allowances by focusing as much on *who should benefit* from the policy as on who should pay. Their accompanying policy proposals captured this new emphasis by not only recommending auctioning allowances but also designating that revenue for programs that would directly benefit the public. This combination of a new public benefit frame and closely matching policy design created a new public benefit model for climate policy—one that has remained influential beyond the RGGI case.

Since RGGI, other climate change policies relying on a similar model have had surprising political success given the increasingly hostile political environment for any climate policies in many places since 2008. This suggests that the public benefit model may represent the best option for enacting new climate policies in the face of growing opposition. Rather than being on the decline, cap and trade may in fact be one of the most promising approaches to enacting a new climate policy, if designed in accordance with the new public benefit framing.

Much of the success of such policies, however, will hinge on the choice of how to allocate revenue or other benefits from the program. In general, policies dedicating auction revenue to tangible public benefits have fared better politically than those seeking to use that revenue for a wider range of government purposes. Post-RGGI policies have also expanded the definition of public benefit to include investments in a diverse set of climate change mitigation and adaptation programs, as well as public health gains from reduced pollutants associated with GHGs. In this respect, the expansion of the public benefit model suggests that the core challenge for future climate change policies may become less about convincing the public of the science of climate change, and more about convincing people that society can address the problem in a fair and equitable manner. The success of the new public benefit model could also support other policies using a price mechanism to regulate pollution as well as those granting private access to other natural resources, such as fisheries, rangelands, or forests.

More broadly, the shift to auctioning allowances indicates the power of norms to create sudden policy change. The success of policies built on the public benefit model in RGGI, the European Union, and California illustrates how policy designs consistent with widely held norms, including egalitarian ideas of fairness central to a "commonsense conception of justice" (Elster 1992), are more politically viable than policies featuring a weaker normative foundation. The failure of recent climate policies that deviated from the public benefit model, such as the Waxman-Markey bill in the United States or Australia's Carbon Pollution Reduction Scheme, also suggests the importance of these fairness norms in separating successful from unsuccessful climate policies across a range of political contexts.

In this manner, the political success of cap and trade with auction in RGGI and other locations offers an example of how the strategy of normative reframing—recasting an issue in terms of alternative norms in order to support a new policy approach—can make dramatic, nonincremental policy change possible. The normative reframing perspective also improves the ability to make predictions about policy stability and change beyond the example of emissions trading, based on variations in public support for normative ideas central to many policy conflicts.

This chapter reviews the theory of normative reframing that is at the heart of the book's explanation for sudden policy change and the evidence of normative reframing's role in the unprecedented policy shift toward auctioning emissions allowances from 2008 to 2015. It then offers some concluding thoughts on the wider implications of the expansion of auctions since RGGI, and the theory of normative reframing for climate and environmental policy making as well as our understanding of sudden policy change in general.

Normative Reframing: An Important Strategy for Policy Change

The core theoretical claim of this book is that norms are a vital factor in creating policy stability and change, and that theories of the policy process would benefit from greater attention to these shared standards of appropriate behavior. Although leading theories of policy making discuss norms in a variety of ways, they tend to focus on the power of norms to create policy *stability* rather than *change*. According to one leading theory, for instance, norms are part of the glue that holds together political coalitions, making policy change more difficult when opposing coalitions share different norms (Jenkins-Smith et al. 2014).

Although some theories hint at the power of norms to create policy change, those accounts remain underdeveloped. John Kingdon's (2003) multiple streams theory of policy change, for example, emphasizes the significance of a policy's "value acceptability" in terms of its potential for enactment, but little subsequent work has tried to depict a detailed norm-based mechanism for policy change within this framework. Punctuated equilibrium theory comes closest to articulating a norm-based account of policy change, describing how policy "punctuations" are often created by new policy images that reconceptualize an issue in ways that could include new normative perspectives (Baumgartner, Jones, and Mortensen 2014). But even this theory says relatively little about the distinctive role of *norms* in promoting such new images, or how one might know when a *particular* policy image is more or less likely to succeed at catalyzing policy change.

This book has argued that it is possible to better explain sudden policy change by concentrating more on the influence of norms. It laid out a theory of normative reframing, in which advocates identify the weak normative underpinnings of a current policy, highlight the weakness of the normative frame supporting the status quo, and offer an alternative frame that recasts the issue in terms of alternative norm(s) that advocates expect the public will perceive as being more applicable to the issue. By reframing the issue in this manner, advocates expect to generate greater support for an alternative policy supported by the new normative frame.

The idea that issue frames affect policy attitudes is uncontroversial. Advocates have long stressed different aspects of a given policy with different issue frames in order to influence public support (Entman 1993). In many cases, it is also understood that issue frames focus on alternative norms—such as describing the decision to permit a hate speech rally in terms of the importance of protecting free speech versus protecting public order (Nelson, Clawson, and Oxley 1997). More attention to the distinctive influence of these normative frames is the key to the theory of policy change portrayed here: by reframing an issue in terms of an alternative norm, advocates can make large policy changes possible even in the face of strong opposition by vested interests.

Norms influence behavior and attitudes through a variety of mechanisms, including fear of sanctions and a desire to maintain a particular identity, and vary in their influence across cultures and specific contexts. Some norms are more central to a person's identity than others, giving those norms a stronger influence over attitudes and behaviors. For example, a norm related to helping others in need is fundamental to the personal identity of many individuals, who offer charity and other forms of assistance even in the absence of social enforcement of the norm. Some norms also generate more serious sanctions for violations than others, making them stronger. Violating certain norms of reciprocity related to exchanges of favors, say, may only result in mild punishment in the form of negative gossip (Ellickson 1991), while other norm violations related to the overuse of an important shared resource like a grazing pasture or fishery may elicit more serious punishments, including ostracism or even exclusion from the community (Ostrom 1990). In this respect, norms vary in their normative force, or degree of influence over behavior and attitudes.

Norms may be ambiguous in their applicability to a given case; individuals often struggle to decide what norm to use in a given situation. Based on this ambiguity, individuals sometimes apply an existing norm to a new situation without much reflection, creating an opportunity for others to argue that the norm creates logical inconsistencies or undesirable consequences in this new circumstance. On reflection, individuals may come to view the norm as offering a poor fit with the situation, making it vulnerable to replacement with an alternative norm.

Even norms that are seen as quite forceful in some circumstances may lose influence when extended to situations where they are later thought to be poorly applied. Norms favoring free expression, for example, may be seen as a poor fit for forms of communication that seriously threaten public safety, and norms favoring equal shares of a resource may be thought to fit poorly with the need to distribute basic goods to families with greatly varying needs after a natural disaster. As with a norm's forcefulness, typical perceptions of a norm's applicability will vary across societies and evolve over time. Nevertheless, most norms are sufficiently well specified that the range of norms that can plausibly be applied to a particular situation or issue is limited.

Assessments of these ideas of normative force and fit are central to the normative reframing process. As noted in chapter 2, normative reframing entails three key steps (figure 6.1). First, change advocates must *identify* the normative frame supporting the current policy—that is, the norm(s) that implicitly or explicitly justify the policy status quo. This normative frame may be obvious in the political discourse surrounding the issue, or it may be more implicit. Either way, advocates must determine what the basic normative justification is for the existing policy.

Having identified the normative frame supporting the current policy, advocates then *foreground* it, trying to highlight either the relative weakness of the norm involved or its weak fit with the issue. In emphasizing a norm's relative weakness, advocates may point out common examples of the norm's violation or argue the negative implications of the norm's application in terms of other valued outcomes. Alternatively, advocates might underscore the poor fit of the norm with the issue, stressing aspects of the new context that make the norm seem less applicable.

By foregrounding the current normative frame as well as highlighting problems of normative force or fit, advocates weaken the normative

1) **Identify normative frame** supporting current policy

2) **Foreground existing normative frame**, highlighting weakness of norm(s) involved or poor fit with the issue, to destabilize policy status quo

3) **Promote new normative frame**, describing the issue in terms of alternative norm(s) with better fit for issue, implying new policy design

Figure 6.1
Key steps in normative reframing strategy.

foundation of the current policy, creating an opportunity to *promote an alternative frame* using a different norm to recommend a different policy solution. In promoting a new normative perspective, advocates reframe the issue in terms of alternative norms they believe will be seen to fit with the issue more closely (or that are simply stronger in that society overall). Advocates can then justify alternative policy approaches in terms of the new norm(s) in the alternative frame.

Normative Reframing and Changes in Cap-and-Trade Policy

Although economists promoting the creation of limited and tradable emissions rights argued for auctioning those rights for decades on the basis of improving the policy's economic efficiency (e.g., Dales 1968), their justifications had little influence on the political process. Powerful interests maintained their access to this important resource even as new cap-and-trade policies began to be adopted, ensuring that those policies gave emissions rights to existing polluters at little or no cost, generally in proportion to their existing levels of emissions (table 6.1). This practice of grandfathering emissions rights was consistent with standard theories of interest group politics, which predict that those with concentrated benefits or costs at stake will exercise greater influence over the final policy design than those with more diffuse interests. In addition, grandfathering persisted for decades because regulators and environmentalists were relatively reluctant to embrace emissions trading and tended not to focus on the allocation issue but rather to default to the politically familiar idea of grandfathering

Table 6.1
New versus Old Model of Cap and Trade

Old model	New model
• Regulators and environmentalists reluctant to use emissions trading; pay little attention to allocation	• Regulators and environmentalists more comfortable with emissions trading; pay more attention to allocation
• Dominance of efficiency-based arguments for auctions	• Emergence of new public benefit frame for auctions
• Greater control of allocation rules by large emitters with large economic interests at stake	• Less control of allocation rules by large emitters in face of new normative frame promoting auctions
• Default to Lockean norm for allocation through grandfathering	• Plans to auction many or all allowances, and dedicate revenue to public benefits

those rights, consistent with Lockean norms for other policies allocating natural resources.

Starting in the 1990s, new developments represented potential seeds of change for the old cap-and-trade model. The 1990 ARP, for example, moved incrementally away from grandfathering toward an alternative allocation that only awarded free allowances up to a certain benchmark emissions rate per unit of fuel consumed. This benchmarking allocation resulted in far fewer allowances for many of the largest polluters compared to what they would have received under a pure grandfathering system. Many states also began to make utility customers pay new public benefit charges on their electricity bills in this period in order to fund programs for energy efficiency in newly deregulated electricity markets. These charges were another departure from the status quo that provided an important precedent for the much larger "taxes" on electricity rates in the form of auctioning emissions allowances in the following decade. Other precedents included experiments with auctioning other public resources, such as portions of the broadcast spectrum, and growing interest in severance taxes on natural resource development.

Despite these seeds of change, emission allowance auctions remained virtually unheard of before RGGI. New cap-and-trade programs for air pollutants such as NO_x in the eastern United States and GHGs in the European Union did experiment with new allocation rules, and featured arguments in favor of auctioning allowances based primarily on polluter pays norms.

Nevertheless, efforts to promote auctioning a significant number of allowances failed in both programs. A careful comparison of both cases with RGGI suggests a key difference: the shift from a simpler normative frame only stressing that the polluter should pay to the more sophisticated frame in RGGI also highlighting the importance of dedicating that revenue to broad public benefits. Unsuccessful auction proposals in the NO_x program and the EU ETS failed to emphasize this new public benefit framing, focusing primarily on the polluter pays justification. Both cases were otherwise similar to RGGI in many ways, further suggesting the importance of the normative framing as a causal factor that made auctioning possible.

As cap-and-trade programs became more familiar, environmental advocates and regulators began to think more seriously about different allocation ideas. Key individuals working on RGGI also identified the importance of having revenue from the sale of emissions allowances benefit the public at large. By adding this egalitarian norm to existing polluter pays framing, environmental advocates were able to generate greater political support for the idea of auctioning and break the control of interest groups over allocation policy, leading to the auction of virtually all allowances under RGGI.

Previous accounts of RGGI's adoption of auctions are grounded in traditional theories of interest group politics, and miss several crucial elements of this "new model" of cap and trade (table 6.1). Contrary to some previous accounts, this book describes how environmental advocates were the pivotal policy entrepreneurs in the adoption of auctions, introducing the idea into the design discussion from the beginning and persuading state agency personnel to support the proposal to sell allowances. In addition, a detailed review of the RGGI design process confirms that the greatest conflicts over auctioning happened relatively early in the process, culminating in the December 2005 signing by seven RGGI states of the MOU that dedicated at least 25 percent of allowances to some form of consumer benefit. After the MOU, momentum for auctioning snowballed, leading states to announce commitments to auction all or nearly all allowances throughout 2006 and early 2007, and dedicate those revenues primarily to consumers. Moreover, the book has shown the political importance of the new public benefit frame portraying auctions as mechanisms for protecting the public from potential energy price increases. Contrary to theories of interest group politics, these arguments for creating broadly distributed consumer benefits were more significant than the provision of free allowances to a few industries in making RGGI possible.

Auction advocates followed the normative reframing strategy outlined above in RGGI. First, they highlighted the weak fit of the Lockean norm justifying private ownership of a public resource based on prior beneficial use. "There is no justification," said one group early in the process, "for continuing to allow 'incumbent' emitters to have a greater right to pollute than others" (MA Climate Coalition 2004). Having stressed the inappropriateness of distributing allowances based on Lockean norms, environmentalists then promulgated their alternative normative frame, combining the polluter pays norm with an egalitarian norm regarding how to distribute revenue from the use of this public resource. As another advocate noted, "We need to show that this program addresses the very serious problem of global warming and pollution but does it in a way that produces a benefit people can get their arms around" (RGGI Stakeholder Meeting Summary, 2005c, 29).

By consciously promoting a new normative frame portraying the atmosphere as a public resource to be used only for broad public benefits, a small group of environmental advocates was able to change political attitudes toward the old approach to allocation. The result was a sudden and remarkable shift toward a new normal where governments own resources like the atmosphere *on behalf of the public*, meaning they should charge firms or individuals for use of this resource, and dedicate those funds to broadly distributed, tangible benefits for all citizens. In this manner, the new public benefit frame built directly on the success of existing and popular public benefit programs to promote energy efficiency, using them as a model for future programs to protect consumers from higher energy prices. Although concerns about problems in the EU ETS were also present, they generally entered the RGGI conversation after the key decisions to auction allowances had been made.

In short, RGGI's political leaders chose to protect the diffuse interests of the public over the concentrated interests of large companies, contrary to the predictions of policy theories grounded in ideas of interest group pluralism. Although fragmentation of the business interests facing the prospect of having to pay for their emissions was part of the explanation for this reversal of the typical pattern, it is not the entire story. Rather, a new normative reframing of the issue based on the idea of requiring that the public benefit directly from any private use of a public resource was pivotal to making auctions a reality in RGGI.

Normative Reframing and RGGI's Policy Implications

Several broader policy implications emerge from a careful review of normative framing and the public benefit model in recent climate policy. First, the public benefit model, including cap and trade with auction, remains one of the more promising approaches for addressing climate change even in light of greater opposition to any climate policy action. Second, continued experiments with different types of public benefits in future frames and policy designs will help define the limits of this new policy approach. Third, the RGGI experience and subsequent events indicate the growing importance of distributional provisions in climate policies compared to efforts to increase public understanding of climate change science. And finally, normative reframing offers a potential to generate policy change in the face of major opposition by vested interests on a wide range of issues beyond climate change.

The Political Potential of the Public Benefit Model

Although relatively little time has passed since RGGI's enactment, ensuing events suggest that the policy represents an inflection point in climate change policy design. Where the old model of cap and trade assumed a free distribution of allowances to emitters, after RGGI the assumption has frequently been that allowances will be auctioned unless there are good reasons to give them away for free. As one cap-and-trade expert puts it, RGGI "shifted the burden of proof" in future emissions trading programs to those seeking free allowances (Tietenberg 2010).

It is important to remember that the movement toward auctions has persevered even in the face of growing political opposition to *any* policies to reduce GHG emissions. California and the European Union are now auctioning a majority of their emissions allowances, and dedicate that revenue to both consumer rebates (in California) as well as diverse programs for climate change mitigation and adaptation (in both California and the EU). Ideas from the public benefit model first developed in RGGI are prominent in both cases, indicating the model's potential to increase support for climate policies in other settings.

The public benefit model has also remained central to RGGI. Organized campaigns attempted to force several RGGI states to withdraw from the compact after 2008, with New Jersey leaving the program in 2011. Some RGGI states also tried to divert auction revenue from energy efficiency and ratepayer assistance programs into their general funds, although with only limited success. Confronted by these challenges, RGGI supporters reemphasized the public benefits of auction revenue, decrying the efforts to divert those funds as violating the heart of the policy. Public officials responded to proposals to withdraw from RGGI by promoting the public benefits of auction revenue investments for their citizens. For this reason, it is not surprising that RGGI's 2013 program review doubled down on auctions, significantly strengthening the regional emissions cap while renewing the commitment to what it now referred to as a cap-and-invest approach.

Meanwhile, recent cap-and-trade policy failures deviated from the public benefit model. In Australia, the eventual repeal of a national carbon pricing policy was driven in part by a failure to sufficiently address public concerns about higher energy prices through the policy's design. In the United States, conservative political groups defeated the major effort to enact ACES (the American Clean Energy and Security Act), a 2010 federal GHG mitigation policy, by contending that the policy would devastate average citizens by creating higher energy prices during a recession. Lacking the argument used by RGGI's advocates that auction revenue would benefit all citizens tangibly by lowering their energy bills, supporters of ACES struggled to respond to this criticism.

Some commentators have interpreted this failure of ACES, a proposal that would have auctioned a majority of its allowances, as a sign of the limits of cap and trade as a policy option. Others have asserted, however, that the failure to dedicate auction revenue to programs that would directly benefit the public was the key reason ACES failed (Skocpol 2013). Former New York governor Pataki (2014), for example, believes that policies such as ACES "damaged the country's ability to deal with this issue in a fair and appropriate manner" by straying from the RGGI model of using auction revenue specifically to protect consumers. Here it is worth noting that new bills have continued to emerge in the US Congress that emphasize returning all revenue from payments for GHG emissions to the public at large (e.g., Nuccitelli 2014).

Defining which Policies Are Consistent with Public Benefit Framing

Even as the public benefit model gains political influence, it is being applied to a wider range of policy designs. Understanding which of those designs can convincingly be made to "fit" with the underlying polluter pays and egalitarian norms in public benefit framing is another important future policy question.

Much of this debate focuses on the appropriate uses of auction revenue for public benefits. This conflict is particularly intense in California, where it has threatened the political viability of AB 32, the 2006 law setting the state's GHG emissions reduction goals. California imposed fewer restrictions than many RGGI states on the uses of auction revenue from the sale of emissions allowances, leading to a smorgasbord of spending proposals and plans. On the one hand, the CPUC has required many electricity suppliers to auction the bulk of their allowances and give those funds to ratepayers in the form of an equal cash dividend. This policy design is largely consistent with the RGGI interpretation of the public benefit frame emphasizing *consumer benefits*, although it focuses on cash rebates rather than subsidies for energy efficiency.

On the other hand, a growing percentage of allowances under the law are auctioned directly by the state, with revenue available for a range of uses consistent with the broad goals of AB 32. Despite two subsequent bills designed to better specify the legitimate uses of this state auction revenue, political conflict continues over the disposition of these funds. Proposals have ranged from support for improved transit options for local communities or a focus on helping disadvantaged communities cope with climate change, to programs aimed more indirectly at reducing emissions such as funding a controversial new high-speed train from San Francisco to Los Angeles. As California expands the definition of public benefits to include these *climate protection benefits*, media observers worry that some of these proposed expenditures are too far removed from the immediate concerns of the law, again raising the problem of "RGGI-cide"—diverting auction revenue to uses that fail to deliver more tangible public benefits related to climate change and energy prices.

Recent efforts to promote the new EPA Clean Power Plan push the boundaries of the public benefit frame even further, attempting to justify a wider range of potential climate policies based on the broadly distributed *public health benefits* created by reducing copollutants associated with

GHGs. In this case, an even wider range of policy designs can be justified by the public benefit frame, including non-market-based policies that do not rely on allowance auctions or pollution taxes. Whether this version of public benefit framing will be as politically effective as those stressing consumer protections or climate change mitigation is an important future question for defining the scope of the public benefit frame.

These debates over defining and distributing the public benefits of a climate policy evoke a long-standing legal and political conflict regarding who is truly the owner of public resources: citizens, or the government that represents them. For example, it is possible to place recent approaches to allocating auction revenue from GHG emissions allowances into two categories based on this conflict. One group of policies treats allowance revenue more like general tax income, giving government broad flexibility in deciding how to spend that money on a variety of public goals. US proposals such as ACES took this approach, and a few RGGI states moved tentatively in this direction by temporarily diverting allowance revenues to the general fund. These approaches treat the atmospheric commons as a public asset *owned primarily by the government* as opposed to the citizens, giving public officials greater latitude in deciding how to allocate the value of this resource.

A second approach, by contrast, treats allowance revenue as part of the *public trust*—resources stewarded by public officials on behalf of their true citizen owners. Governments are not free to dedicate benefits from the use of public trust resources such as the atmospheric commons to a wide variety of purposes. Instead, governments hold those resources in trust for the public, and must ensure that most or all of the public benefits directly from any policy affecting that resource. This approach is quite similar to the long-standing public trust doctrine in the common law that prevents the government from granting private rights to waterways or other resources in ways that limit the public's access or ability to benefit from those resources (Sax 1970; Wood 2014). Under this conception, the atmospheric commons is *owned primarily by the citizens*, rather than their government representatives.

Although it is still early to be drawing conclusions, the emerging pattern seems to be that programs following the narrower public trust definition of public benefits fare better politically. This suggests that future cap-and-trade programs would do well to consider dedicating auction revenue directly to broad public benefits that are consistent with the public trust conception of the atmospheric commons and RGGI's public benefit framing. At the same time, the ongoing debate in California indicates that effective approaches to defining the appropriate public benefit from auction revenue or a carbon tax may vary, potentially including direct rebates, subsidies for programs to reduce energy consumption, or other strategies benefiting most or all citizens as directly as possible by providing broad climate protection benefits. More recent efforts to expand the scope of public benefits to include gains in public health from reduced emissions of pollutants associated with GHGs is another important test of the range of policies that will be perceived as fitting with the new public benefit frame, including potential provisions for ensuring more equal access to healthy air consistent with the demands of environmental justice advocates.

The Importance of Policy Design versus Scientific Communication

Many continue to argue that the key to advancing more ambitious climate change policies is to explain climate science more effectively, increasing public understanding of and concern about the issue, and thereby pressure for policy action. The success of the public benefit model, by contrast, suggests that climate policy progress may depend more on finding and promoting the right distribution of costs and benefits from any proposed solution rather than increasing belief in climate science.

Although public understanding of and concern about climate change will always remain an important political consideration, it is worth noting that climate policies continue to founder even in locations where a majority of the public expresses belief in human-induced climate change and worry about the impacts of those changes (Krosnick and MacInnis 2011; Nisbet and Myers 2007). In addition, psychological research suggests that skepticism about climate science can be the consequence rather than the cause of climate policy attitudes. When presented with free market or technological solutions to climate change, for example, politically conservative individuals tended to be more accepting of the idea of human contributions to climate change than when presented with traditional regulatory approaches (Campbell and Kay 2014). This is consistent with other research indicating that political identities are a critical predictor of agreement with climate science, even among well-educated individuals (Kahan 2010, 2015).

The events described in this book are consistent with the greater significance of policy design choices over questions of science communication in determining the political fate of climate policies. The conflict over Australia's 2012 law reducing GHG emissions through a cap-and-trade approach, for instance, did not hinge on skepticism about climate change; both sides conceded the need for action on the issue. Instead, the conflict was primarily over who should bear the costs of those actions, and how policies might best address climate change without hurting average citizens economically.

Similar dynamics continue to play out in the United States. In a 2014 speech to environmental groups, President Obama spoke bluntly about the importance of any climate policy's economic impacts: "People don't like gas prices going up; they are concerned about electricity prices going up. If we're blithe about saying, 'This is the crisis of our time,' but we don't acknowledge these legitimate concerns—we've got to shape our strategies to address the very real and legitimate concerns of working families" (quoted in Davenport 2014). As mentioned in chapter 5, members of the Obama administration are making similar arguments that the EPA's CPP regulations can reduce consumers' energy bills as well as GHG emissions, by funding investments in reducing energy consumption. Meanwhile, protests from those opposed to these climate policies continue to focus on potentially high costs for consumers and the economy.

The policy controversies described in this book, in short, testify to the growing importance of determining who will pay and who will benefit from any GHG mitigation policy compared to concerns over convincing the public that climate change is real. Although some of the public remains skeptical of climate science, such voices are a minority even in the United States, where a majority of citizens are convinced that the Earth is warming (Dimock et al. 2014; Leiserowitz et al. 2014). Given these results along with evidence that policy attitudes can influence scientific beliefs about climate change, it seems clear that we have entered an era where finding the right policy design to reduce GHG emissions at the lowest cost to the public at large will be at least as important as finding new ways to persuade more members of the public about the validity of climate science.

Normative Reframing in Other Major Policy Changes

Finally, normative reframing represents a strategy for creating major policy change in the face of opposition from powerful interests on a wide range of issues beyond climate change. Prior to RGGI, few would have predicted that environmental advocates would be able to force large economic actors to pay for their emissions in the form of an allowance auction or emissions tax. Large polluters had successfully prevented the serious consideration of emissions charges for decades, with few signs that their influence over this process was likely to change in the future. Despite a few small modifications to this old model of cap and trade in the 1990s, no effort to make power generators pay for their existing level of emissions was close to being successful until RGGI.

RGGI's sudden reversal of that trend suggests that change advocates can succeed in generating major policy reforms even when up against opposition from powerful interests by identifying and targeting policies with weak normative foundations, and offering stronger and more convincing normative frames for their preferred alternative. By drawing on the power of norms to shape many human attitudes and behaviors, advocates can overcome resistance to changes in policies serving vested interests by generating greater public support for an alternative that fits better with powerful social norms applied to the issue.

Chapter 5 described how normative reframing could apply to many other policy issues. Any policy grounded on a normative foundation that relies on relatively weak norms or norms that are not persuasively applied to an issue should be vulnerable to the same kind of change strategy. The breadth of policy areas subject to important normative considerations, ranging from distributive policies for various public goods including transportation or health care, to policies governing hot-button social issues such as gay marriage, abortion, or civil rights, points to the broad potential application of normative reframing as a policy change strategy. Advocates have used norm-based strategies to generate policy change in the past, both by weakening existing norms (Weldon 2006) and reframing issues in terms of alternative norms that recommend different policy approaches. Although these efforts have been documented in some specific areas of social and moral policy, including civil rights reforms, the broader potential of normative reframing as a conceptual strategy has not been recognized. This book asserts that the realm of policies subject to norm-based strategies for change is actually quite large, including many environmental and other policies where the distribution of public goods is a key part of the policy design question.

Normative Reframing and RGGI's Theoretical Implications

Most theories of the policy process are pessimistic about the ability to predict specific policy changes, tending instead to describe the many forces that create policy stability. Where theories do offer predictions of more dramatic policy changes, they limit themselves to depicting the expected distribution of such changes over a period of time rather than trying to identify factors that indicate a particular policy is ripe for change. This book has argued that these theories are too pessimistic and ignore what change advocates do in practice on a regular basis: identify policies that are especially vulnerable to reform efforts.

One important way that activists can identify such policies is by looking for weak normative foundations—normative frames used to justify the policy that rely on relatively weak norms in that society or risk being perceived by the public as a poor fit with the issue. As noted by many existing policy theories, norms play a central role in shaping human attitudes and behaviors. Studying a specific policy's normative foundation is a promising approach to identifying if it is more likely to experience sudden policy change: the weaker that foundation, the more likely sudden change could happen.

This new approach to predicting policies that are most vulnerable to change does not require major new methodological innovations. Scholars and advocates already identify the major normative frames supporting a given policy by investigating the rhetoric and discourse surrounding the policy's adoption and subsequent implementation. This can be done both by interviewing advocates on all sides of the issue regarding the key normative conflicts in a policy debate as well as analyzing how frequently different norms appear in records of public and private discussions of the issue, or even in political discourse in general. Norms cited by advocates as being most central to their policy arguments, or that appear most frequently or at the most critical moments in those policy debates, are likely to form the current policy's normative foundation.

Having established the normative frame or frames that support the current policy, the next step is to determine how strong those norms are in society and how well they appear to fit the issue at hand in the eyes of elites or the public at large. Here again, no new methods are required; scholars already investigate the relative strength or force of different norms over individual attitudes and behaviors in a variety of ways, relying on observations and historical accounts of costly behavior in daily life, or on behavior in laboratory experiments consistent with certain normative principles, along with surveys and interviews about the perceived significance of different norms in different decision contexts.

Although studies of a norm's perceived fit with an issue are less common, they are subject to many of the same techniques. Standard survey questions that are already used to test the strength and persuasiveness of different frames for a given issue could easily be applied to ask respondents how relevant a particular set of norms is to a given issue. Researchers have also used interviews to explore individual perceptions of normative fit, and have investigated those perceptions experimentally by observing behavioral deviations from a norm under different conditions in the lab or field. Finally, empirical work has documented how individuals modify distributional norms (for example, equal shares for everyone) to apply to different types of distributive situations such as organ donation or school admissions (Elster 1992, 1995).

By identifying policies where the dominant normative frame is perceived as weak by the public, either due to the weakness of the norm being cited or its apparent lack of relevance to the issue at hand, scholars and advocates should be able to predict policies that are more or less vulnerable to nonincremental change. This would be a useful step forward for most theories of the policy process, which currently offer few hypotheses for predicting sudden policy changes.

Of course, not all policies that are built on a weak normative foundation will change in the near future; there must be motivated change agents who are willing to mobilize and push for policy reform, offering new normative frames to replace the currently weak one. Without these agents, policies with weak normative frames will persist. The strong influence of powerful interest groups as well as structural and formal rules that create a tendency toward the status quo in many political systems also remain in effect even for policies with a weak normative foundation. In some cases, those forces will win out even in the face of a reform effort based on a normative reframing strategy.

The importance of normative reframing, then, is not that it offers a guaranteed method of sudden policy change. Rather, it offers a promising strategy for overcoming the strong institutional forces that resist many policy changes, drawing on the unique power of norms to surmount those forces opposed to change. By identifying and foregrounding the normative weakness of the frames supporting the status quo, and offering an alternative policy grounded in a more persuasive normative frame, advocates can make policy change more likely even in the face of resistance from powerful interests. This is exactly what auction supporters did in the RGGI case, offering a new and more compelling public benefit frame to justify selling emissions allowances that went beyond previous efforts to make polluters pay for their resource use. It is also a pattern that seems to have occurred or be occurring in other cases, such as promoting gay marriage in the first decades of the twenty-first century.

For these reasons, change advocates and policy theorists would do well to pay closer attention to the lessons of the RGGI revolution. The strategy of normative reframing offers a promising way to identify policies with weak normative foundations that are more vulnerable to sudden change and a method for promoting policy change even in the face of strong resistance from vested interests. By foregrounding the weakness of the normative frame supporting a policy, advocates can put supporters of the status quo in a difficult political position. By offering a more compelling alternative frame that supports a different policy approach, advocates can sometimes make previously unthinkable policy ideas a political reality.

Complete List of Interviewees

Ian Bowles, (formerly) MA Office of Energy and Environmental Affairs (secretary)

Michael Bradley, Northeast Regional Greenhouse Gas Coalition and MJ Bradley Group

Dwayne Breger, MA Department of Energy Resources (member RGGI SWG)

Marc Breslow, (formerly) MA Climate Coalition

Dale Bryk, National Resources Defense Council

Robert Burnley, (formerly) VA Department of Environmental Quality

Dallas Burtraw, Resources for the Future

Bradley Campbell, (formerly) NJ Department of Environmental Protection (commissioner)

David Cash, (formerly) MA Office of Energy and Environmental Affairs (undersecretary for policy)

Richard Cowart, Regulatory Assistance Project

Laurence DeWitt, Pace Law School Energy and Climate Center

Douglas Foy, (formerly) MA Department of Environmental Protection

Sonia Hamel, (formerly) MA Office of Commonwealth Development (cochair RGGI SWG)

Seth Kaplan, (formerly) Conservation Law Foundation

Jesse Kharbanda, Hoosier Environmental Council

Joseph Kwasnik, (formerly) National Grid Utility

William Lamkin, MA Department of Environmental Protection (member RGGI SWG)

Franz Litz, (formerly) NY Department of Environmental Conservation (cochair, RGGI SWG)

Mary Major, VA Department of Environmental Quality

Gina McCarthy, (formerly) CT Department of Environmental Protection (commissioner)

Neal Menkes, VA State Senate Finance Committee (staff member)

William Murray, (formerly) VA Governor's Office (deputy policy director)

Derek Murrow, (formerly) Environment Northeast

Marc Pacheco, MA state senator

George Pataki, (formerly) NY governor

David Paylor, (formerly) VA Department of Natural Resources (deputy secretary)

Robert Rio, Associated Industries of Massachusetts

Robert Ruddock, (formerly) Associated Industries of Massachusetts

Nancy Seidman, MA Department of Environmental Protection

Denise Sheehan, (formerly) NY Department of Environmental Conservation (head)

William Shobe, (formerly) VA Department of Planning and Budget

Rob Sliwinski, NY Department of Environmental Conservation

Jessie Stratton, (formerly) Environment Northeast

Eric Svenson, (formerly) Public Service Enterprise Group Utility

Mark Sylvia, (formerly) MA Office of Energy and Environmental Affairs (undersecretary for energy)

References

Abbott, Tony. 2010. Direct Action on the Environment and Climate Change. Press release, Office of Honorable Tony Abbott, February 2. http://parlinfo.aph.gov.au/parlInfo/search/display/display.w3p;query=Id%3A%22media%2Fpressrel%2FGMSV6%22 (accessed December 5, 2015).

Ackerman, Bruce A., and Richard B. Stewart. 1985. Reforming Environmental Law. *Stanford Law Review* 37:1333–1365.

Adirondack Council. 2006. Comments on RGGI Draft Model Rule, May 23. http://www.rggi.org/design/history/public_comments (accessed November 28, 2015).

AES. 2004. Public Comment, Assessment of Public Benefits Set-Aside Concept, October 11. http://www.rggi.org/design/history/public_comments (accessed November 25, 2015).

AK Department of Revenue. 2014. Historical Summary of Dividend Applications and Payments. http://pfd.alaska.gov/Division-Info/Summary-of-Applications-and-Payments (accessed September 24, 2015).

American Council for an Energy Efficient Economy (ACEEE) and Alliance to Save Energy. 2006. Public Comment on RGGI Draft Model Rule, May 22. http://www.rggi.org/design/history/public_comments (accessed November 28, 2015).

American Electric Power. 2006. Public Comments on RGGI Draft Model Rule, May 18. http://www.rggi.org/design/history/public_comments (accessed November 28, 2015).

Appalachian Mountain Club. 2006. Comments on RGGI Draft Model Rule, May 23. http://www.rggi.org/design/history/public_comments (accessed November 28, 2015).

Ariely, Dan. 2009. *Predictably Irrational*. New York: HarperCollins.

Arnold, R. Douglas. 1990. *The Logic of Congressional Action*. New Haven, CT: Yale University Press.

Arup, Tom. 2015. "Carbon Tax Zombies? Direct Action, Emissions Trading, and the Carbon Tax Explained." *Sydney Morning Herald*, July 15. http://www.smh.com.au/federal-politics/political-news/carbon-tax-zombies-direct-action-emissions-trading-and-the-carbon-tax-explained-20141030-11ebgx (accessed December 5, 2015).

Aulisi, Andrew, Alexander E. Farrell, Jonathan Pershing, and Stacy Vandeveer. 2005. *Greenhouse Gas Emissions Trading in U.S. States: Observations and Lessons from the OTC NO$_x$ Budget Program*. Washington, DC: World Resources Institute.

Australian House of Representatives. 2009. Carbon Pollution Reduction Scheme Bill 2009. http://parlinfo.aph.gov.au/parlInfo/search/display/display.w3p;query=Id%3A%22legislation%2Fbillhome%2Fr4127%22#ems (accessed December 5, 2015).

Australian Senate. 2011. Clean Energy Act 2011, Revised Explanatory Memo. http://parlinfo.aph.gov.au/parlInfo/search/display/display.w3p;query%3DId%3A%22legislation%2Fbillhome%2Fr4653%22;rec=0 (accessed December 5, 2015).

Australian Department of Climate Change. 2008. Carbon Pollution Reduction Scheme: Australia's Low Pollution Future. White paper, December 15. http://pandora.nla.gov.au/pan/99543/20090515-1610/www.climatechange.gov.au/whitepaper/report/index.html (accessed December 5, 2015).

Australian Department of Climate Change and Energy Efficiency. 2011. Securing a Clean Energy Future: The Australian Government's Climate Change Plan. On file with author.

Avril, Tom. 2005. "Nine States Ready to Fight Pollution on Their Own." *Philadelphia Inquirer*, December 12.

Axelrod, Robert. 1986. An Evolutionary Approach to Norms. *American Political Science Review* 80 (4): 1095–1111.

Bachram, Heidi. 2004. Climate Fraud and Carbon Colonialism: The New Trade in Greenhouse Gases. *Capitalism, Nature, Socialism* 15:5–20.

Backus, Bob. 2015. "Changes to RGGI Are Penny Wise and Pound Foolish." *New Hampshire Business Review*, March 6. http://www.nhbr.com/March-6-2015/Changes-to-RGGI-are-penny-wise-and-pound-foolish/ (accessed December 1, 2015).

Baker-Branstetter, Shannon. 2014. "Another View: Consumers Can't Afford a Delay in Cap-and-Trade." *Sacramento Bee*, August 24. http://www.sacbee.com/opinion/california-forum/article2607303.html (accessed December 4, 2015).

Bali, Valentina. 2009. Tinkering toward a National Identification System: An Experiment on Policy Attitudes. *Policy Studies Journal* 37 (3): 233–255.

Barnes, Peter. 2001. *Who Owns the Sky?* Washington, DC: Island Press.

Barnes, Peter, and Marc Breslow. 2003. The Sky Trust: The Battle for Atmospheric Scarcity Rent. In *Natural Assets: Democratizing Environmental Ownership*, ed. James K. Boyce and Barry G. Shelley, 135–150. Washington, DC: Island Press.

Baron, Jonathan, and Mark Spranca. 1997. Protected Values. *Organizational Behavior and Human Decision Processes* 70 (1): 1–16.

Barringer, Felicity, and Kate Galbraith. 2008. "States Aim to Cut Gases by Making Polluters Pay." *New York Times*, September 15. http://www.nytimes.com /2008/09/16/us/16carbon.html?pagewanted=all&_r=0 (accessed November 29, 2015).

Baumgartner, Frank R., Suzanna De Boef, and Amber Boydstun. 2008. *The Decline of the Death Penalty and the Discovery of Innocence*. Cambridge: Cambridge University Press.

Baumgartner, Frank R., and Bryan D. Jones. 1993. *Agendas and Instability in American Politics*. Chicago: University of Chicago Press.

Baumgartner, Frank R., Bryan D. Jones, and Peter B. Mortensen. 2014. Punctuated Equilibrium Theory: Explaining Stability and Change in Public Policymaking. In *Theories of the Policy Process*, ed. Paul A. Sabatier and Christopher M. Weible, 59–103. Boulder, CO: Westview Press.

Baumol, William J., and Wallace E. Oates. 1971. The Use of Standards and Prices for Protection of the Environment. *Swedish Journal of Economics* 73 (1): 42–54.

Baunach, Dawn Michelle. 2011. Decomposing Trends in Attitudes toward Gay Marriage, 1988–2006. *Social Science Quarterly* 92 (2): 346–363.

Becker, Julia, and Janet K. Swim. 2012. Seeing the Unseen: Attention to Daily Encounters as a Way to Reduce Sexist Beliefs. *Psychology of Women Quarterly* 35:227–242.

Béland, Daniel, and Robert Henry Cox, eds. 2011. *Ideas and Politics in Social Research*. Oxford: Oxford University Press.

Benford, Robert D., and David A. Snow. 2000. Framing Processes and Social Movements: An Overview and Assessment. *Annual Review of Sociology* 26:611–639.

Berman, Sheri. 1998. *The Social Democratic Moment: Ideas and Politics in the Making of Interwar Europe*. Cambridge, MA: Harvard University Press.

Berman, Sheri. 2001. Ideas, Norms, and Culture in Political Analysis. *Comparative Politics* 33 (2): 231–250.

Bernstein, Marver H. 1955. *Regulating Business by Independent Commission*. Princeton, NJ: Princeton University Press.

Betsill, Michele, and Matthew J. Hoffmann. 2011. The Contours of "Cap and Trade": The Evolution of Emissions Trading Systems for Greenhouse Gases. *Review of Policy Research* 28 (1): 83–106.

Bierbower, Will. 2011. A Brief History of Fraudulent Activity on the EU-ETS. Worldwatch Institute ReVolt! (blog), February 25. http://blogs.worldwatch.org/revolt/a-brief-history-of-fraudulent-activity-on-the-eu-ets-2/ (accessed December 1, 2015).

Binmore, Ken, and Paul Klemperer. 2001. *The Biggest Auction Ever: The Sale of the British 3G Telecom Licences*. Oxford: Oxford University Press.

Blyth, Mark. 2002. *Great Transformations: Economic Ideas and Institutional Change in the Twentieth Century*. Cambridge: Cambridge University Press.

Boling, Patricia. 2015. *The Politics of Work–Family Policies: Comparing Japan, France, Germany, and the United States*. Cambridge: Cambridge University Press.

Bollinger, Mark, Ryan Wiser, Lew Milford, Michael Stoddard, and Kevin Porter. 2001. Clean Energy Funds: An Overview of State Support for Renewable Energy. Lawrence Berkeley National Laboratory Report 47705. https://emp.lbl.gov/publications/clean-energy-funds-overview-state (accessed March 30, 2016).

Borah, Porismita. 2011. Conceptual Issues in Framing Theory: A Systematic Examination of a Decade's Literature. *Journal of Communication* 61:246–263.

Bowles, Ian. 2014. Interview, June 11.

Bowles, Samuel, and Herbert Gintis. 2011. *A Cooperative Species: Human Reciprocity and Its Evolution*. Princeton, NJ: Princeton University Press.

Bradley, Michael. 2005. Talking Points for the April 6, 2005, RGGI Stakeholder Meeting Regarding the RFF RGGI Allocation Study. Presentation to the RGGI Stakeholder Group Meeting, April 6. On file with author.

Bradley, Michael. 2011. Interview, October 14.

Bradley, Michael. 2013. Interview, November 7.

Braine, Bruce. 2004. Comments on Auction vs. Allocation of CO_2 Allowances. Presentation to the RGGI Allocations Workshop, October 14. http://www.rggi.org/docs/braine_pres_10_14_04.pdf (accessed November 25, 2015).

Breger, Dwayne. 2010. Interview, November 12.

Breslow, Marc. 2011. Interview, August 23.

Brewer, Paul R. 2001. Value Words and Lizard Brains: Do Citizens Deliberate about Appeals to Their Core Values? *Political Psychology* 22 (1): 45–64.

Brewer, Paul R. 2003. Values, Political Knowledge, and Public Opinion about Gay Rights. *Public Opinion Quarterly* 67:173–201.

British Columbia Ministry of Finance. 2013. Carbon Tax Review. In *June Budget Update: 2013/14–2015/16*. http://www.bcbudget.gov.bc.ca/2013_june_update/ bfp/2013_June_Budget_Fiscal_Plan.pdf (accessed December 5, 2015).

Broder, John. 2010. "'Cap and Trade' Loses Its Standing as Energy Policy of Choice." *New York Times*, March 25. http://www.nytimes.com/2010/03/26/science/ earth/26climate.html (accessed December 4, 2015).

Bryk, Dale S. 2004. Public Benefit Allowance Allocations. Presentation at the RGGI Allocations Workshop, October 14. http://www.rggi.org/docs/bryk_pres_10_14_04 .pdf (accessed November 25, 2015).

Bryk, Dale. 2011. Interview, April 6.

Bryk, Dale. 2014. Interview, January 21.

Bryner, Gary C. 1993. *Blue Skies, Green Politics: The Clean Air Act of 1990*. Washington, DC: CQ Press.

Burnley, Robert. 2013. Interview, September 11.

Burtraw, Dallas. 2011. Interview, February 18.

Burtraw, Dallas. 2013. Interview, October 11.

Burtraw, Dallas, and David A. Evans. 2003. The Evolution of NO_x Control Policy for Coal-Fired Power Plants in the United States. Resources for the Future Discussion Paper 03–23. http://www.rff.org/research/publications/evolution-nox -control-policy-coal-fired-power-plants-united-states (accessed November 20, 2015).

Burtraw, Dallas, and Karen Palmer. 2003. The Paparazzi Take a Look at a Living Legend: The SO_2 Cap-and-Trade Program for Power Plants in the United States. Resources for the Future Discussion Paper 03–15. http://www.rff.org/research/ publications/paparazzi-take-look-living-legendthe-so2-cap-and-trade-program -power-plants (accessed November 18, 2015).

Burtraw, Dallas, and Karen Palmer. 2004. Initial Allocation of CO_2 Allowances in the Regional Greenhouse Gas Initiative: Preliminary Observations. Presentation at the RGGI Stakeholder Group Meeting, June 24. On file with author.

Burtraw, Dallas, Karen Palmer, and Danny Kahn. 2005. Allocation of CO_2 Emission Allowances in RGGI. Presentation to the RGGI Stakeholder Meeting, April 6. On file with author.

Burtraw, Dallas, and Byron Swift. 1996. A New Standard of Performance: An Analysis of the Clean Air Act's Acid Rain Program. *Environmental Law Reporter* 26 (8): 10411–10423.

Burtraw, Dallas, and Sarah Jo Szambelan. 2009. U.S. Emissions Trading Markets for SO_2 and NO_x. Resources for the Future Discussion Paper 09–40. http://www.rff.org/research/publications/us-emissions-trading-markets-so2-and-nox (accessed November 19, 2015).

Business Council of New York State. 2006. Public Comments on RGGI Draft Model Rule, May 23. http://www.rggi.org/design/history/public_comments (accessed November 28, 2015).

Buying Property with a Shovel. 2010. Freakonomics blog, February 10. http://freakonomics.com/2010/02/10/buying-property-with-a-shovel/ (accessed November 14, 2015).

California Air Resources Board (CARB). 2008. Climate Change Scoping Plan, December 2008. http://www.arb.ca.gov/cc/scopingplan/document/adopted_scoping_plan.pdf (accessed December 4, 2015).

California Air Resources Board (CARB). 2013a. Cap-and-Trade Auction Proceeds Investment Plan: Fiscal Years 2013–14 and 2015–16, May 14. http://www.arb.ca.gov/cc/capandtrade/auctionproceeds/final_investment_plan.pdf (accessed December 4, 2015).

California Air Resources Board (CARB). 2013b. Title 17, California Code of Regulations, Article 5: California Cap on Greenhouse Gas Emissions and Market-Based Compliance Mechanisms to Allow for the Use of Compliance Instruments Issued by Linked Jurisdictions, July. http://www.arb.ca.gov/cc/capandtrade/ctlinkqc.pdf (accessed December 4, 2015).

California Air Resources Board (CARB) Economic and Allocation Advisory Committee. 2010. Allocating Emissions Allowances under a California Cap-and-Trade Program, March 2010. http://www.climatechange.ca.gov/eaac/documents/eaac_reports/2010-03-22_EAAC_Allocation_Report_Final.pdf (accessed December 4, 2015).

California Public Utilities Commission (CPUC). 2012. Decision Adopting Cap-and-Trade Greenhouse Gas Allowance Revenue Allocation Methodology for the Investor-Owned Electric Utilities, Decision 12-12-033, December 28. http://docs.cpuc.ca.gov/PublishedDocs/Published/G000/M040/K631/40631611.pdf (accessed December 4, 2015).

Campbell, Bradley. 2012. Interview, November 13.

Campbell, Troy H., and Aaron C. Kay. 2014. Solution Aversion: On the Relation between Ideology and Motivated Disbelief. *Journal of Personality and Social Psychology* 107 (5): 809–824.

Carlson, Laurel J. 1996. NESCAUM/MARAMA NO_x Budget Model Rule. Providence, RI, January 30.

Carrington, Damian. 2013. "EU Carbon Price Crashes to Record Low." *Guardian*, January 24. http://www.theguardian.com/environment/2013/jan/24/eu-carbon -price-crash-record-low (accessed December 1, 2015).

Cash, David. 2014. Interview, August 5.

Catholic Diocese of Rochester, New York (Public Policy Committee). 2006. Comments on RGGI Draft Model Rule, May 22. http://www.rggi.org/design/history/ public_comments (accessed November 28, 2015).

Center for Clean Air Policy. 1999. Design of a Practical Approach to Greenhouse Gas Emissions Trading Combined with Policies and Measures in the EC, November. http://ccap.org/resource/design-of-a-practical-approach-to-greenhouse-gas-emissions -trading-combined-with-policies-and-measures-in-the-ec/ (accessed November 20, 2015).

Center for Clean Air Policy. 2003. Recommendations to Governor Pataki for Reducing New York State Greenhouse Gas Emissions, April. http://ccap.org/assets/Recommendations-to-Governor-Pataki-for-Reducing-New-York-State-Greenhouse-Gas-Emissions_CCAP_April-2003.pdf (accessed January 16, 2016).

Center for Energy and Economic Development. 2003a. Public Comment—NE CO_2 Analysis (Economic), August. http://www.rggi.org/design/history/public_comments (accessed November 24, 2015).

Center for Energy and Economic Development. 2003b. Public Comment—NE CO_2 Analysis (Environmental), August. http://www.rggi.org/design/history/public _comments (accessed November 24, 2015).

Center for Energy and Economic Development. 2004. Public Comment—Charles River Associates RGGI, Economic Analysis, July 20. http://www.rggi.org/design/history/public_comments (accessed November 25, 2015).

Chaffin, Joshua. 2012. "Emissions Trading: Cheap and Dirty." *Financial Times*, February 14. http://www.ft.com/intl/cms/s/0/135e1172-5636-11e1-8dfa-00144feabdc0 .html#axzz3mzXZn7qE (accessed December 1, 2015).

Chong, Dennis. 2000. *Rational Lives: Norms and Values in Politics and Society*. Chicago: University of Chicago Press.

Chong, Dennis, and James N. Druckman. 2007. Framing Theory. *Annual Review of Political Science* 10:103–126.

Chudek, Maciej, and Joseph Henrich. 2011. Culture-Gene Coevolution, Norm-Psychology, and the Emergence of Human Prosociality. *Trends in Cognitive Sciences* 15 (5): 218–226.

Clawson, Rosalee A., and Rakuya Trice. 2000. Poverty as We Know It: Media Portrayals of the Poor. *Public Opinion Quarterly* 64:53–64.

Clean Water Action. 2006. Comments on RGGI Draft Model Rule, May 22. http://www.rggi.org/design/history/public_comments (accessed November 28, 2015).

Cliff, Steven. 2014. Personal communication, June 12.

Climate Network Europe. 2000. Comments on European Commission Green Paper, October. http://ec.europa.eu/environment/archives/docum/pdf/0087_ngo.pdf (accessed November 20, 2015).

Coase, Ronald H. 1959. The Federal Communications Commission. *Journal of Law and Economics* 2:1–40.

Coase, Ronald H. 1960. The Problem of Social Cost. *Journal of Law and Economics* 3:1–44.

Cohen, Michael D., James G. March, and Johan P. Olsen. 1972. A Garbage Can Model of Organizational Choice. *Administrative Science Quarterly* 17:1–25.

Cohen, Richard E. 1995. *Washington at Work: Back Rooms and Clean Air.* Boston: Allyn and Bacon.

Cole, Daniel H. 2002. *Pollution and Property.* Cambridge: Cambridge University Press.

Collier, David. 2011. Understanding Process Tracing. *PS: Political Science and Politics* 44:823–830.

Collier, David, Henry E. Brady, and Jason Seawright. 2010. Sources of Leverage in Causal Inference: Toward an Alternative View of Methodology. In *Rethinking Social Inquiry*, ed. Henry E. Brady and David Collier, 161–199. Lanham, MD: Rowman and Littlefield.

Conectiv Energy. 2006. Public Comments on RGGI Draft Model Rule, May 23. http://www.rggi.org/design/history/public_comments (accessed November 28, 2015).

"Connecticut Governor Wants Allowances for RGGI CO_2 Program All to Be Auctioned." 2007. *Electric Utility Week*, March 12.

Conrad, Mark A. 1989. The Demise of the Fairness Doctrine: A Blow for Citizen Access. *Federal Communications Law Journal* 41:161–194.

Consumer Power Advocates. 2006. Comments on RGGI Draft Model Rule, May 22. http://www.rggi.org/design/history/public_comments (accessed November 28, 2015).

Convery, Frank J., and Luke Redmond. 2013. The European Union Emissions Trading Scheme: Issues in Allowance Price Support and Linkage. *Annual Review of Resource Economics* 5:301–324.

Cook, Brian J. 2010. Arenas of Power in Climate Change Policymaking. *Policy Studies Journal* 38 (3): 465–486.

Corombos, Greg. 2015. "Inhofe: Obama Reincarnating Failed Cap-and-Trade: New Environmental Plan 'Most Regressive of All Taxes Ever Passed.'" *WND Radio,* August 8. http://www.wnd.com/2015/08/inhofe-obama-reincarnating-failed-cap-and-trade/#Tkjp3S0SV648fCIP.99 (accessed December 5, 2015).

Costanza, Robert. 2015. "Claim the Sky!" *Solutions,* April. http://www.thesolutions journal.com/node/237301 (accessed December 4, 2015).

Cowart, Rich. 2011. Interview, October 17.

Craemer, Thomas. 2009. Framing Reparations. *Policy Studies Journal* 37 (2): 275–298.

Cramton, Peter. 2002. Spectrum Auctions. In *Handbook of Telecommunications Economics,* ed. Martin Cave, Sumit Majumdar, and Ingo Vogelsang, 605–639. Amsterdam: Elsevier Science.

Crandall, Christian, Amy Eshleman, and Laurie O'Brien. 2002. Social Norms and the Expression and Suppression of Prejudice: The Struggle for Internalization. *Journal of Personality and Social Psychology* 82:359–378.

Crocker, Thomas D. 1966. The Structuring of Atmospheric Pollution Control Systems. In *The Economics of Air Pollution,* ed. Harold Wolozin, 61–86. New York: W. W. Norton and Company.

CT Industrial Energy Consumers Coalition. 2006. Public Comments on RGGI Draft Model Rule, May 22. http://www.rggi.org/design/history/public_comments (accessed November 28, 2015).

C2ES. 2014. California Cap-and-Trade Program Summary. Washington, DC: Center for Climate and Energy Solutions. http://www.c2es.org/docUploads/calif-cap-trade -01-14.pdf (accessed December 1, 2015).

Cunningham, Dan. 2004. Untitled. Presentation to the RGGI Allocations Workshop, October 14. http://www.rggi.org/docs/cunningham_pres_10_14_04.pdf (accessed November 25, 2015).

Cushman, John H. 2014. "Why EPA Expects Its New Carbon Rule to Shrink Your Electric Bill." *Inside Climate News,* June 20. http://insideclimatenews.org/carbon -copy/20140620/why-epa-expects-its-new-carbon-rule-shrink-your-electric-bill (accessed December 5, 2015).

Dales, John H. 1968. *Pollution, Property, and Prices.* Toronto: University of Toronto Press.

Danish Energy Agency. 2000. Comments on European Commission Green Paper. November. http://ec.europa.eu/environment/archives/docum/pdf/0087_govern mental.pdf (accessed November 20, 2015).

Dardis, Frank E., Frank R. Baumgartner, Amber E. Boydstun, Suzanna De Boef, and Fuyuan Shen. 2008. Media Framing of Capital Punishment and Its Impact on Individuals' Cognitive Responses. *Mass Communication and Society* 11 (2): 115–140.

Davenport, Coral. 2014. "Climate Campaign Can't Be Deaf to Economic Worries, Obama Warns." *New York Times*, June 26. http://www.nytimes.com/2014/06/26/us/politics/obama-warns-climate-campaign-cant-be-deaf-to-economic-worries.html (accessed December 6, 2015).

Dawes, Christopher T., Magnus Johannesson, Erik Lindquist, Peter Loewen, Robert Ostling, Marianne Bonde, and Frida Priks. 2012. *Generosity and Political Preferences*. Stockholm: Research Institute of Industrial Economics.

Delshad, Ashlie B., and Leigh Raymond. 2013. Media Framing and Public Attitudes toward Biofuels. *Review of Policy Research* 30 (2): 190–210.

Delshad, Ashlie B., Leigh Raymond, Vanessa Sawicki, and Duane T. Wegener. 2010. Public Attitudes toward Political and Technological Options for Biofuels. *Energy Policy* 38:3414–3425.

DeWitt, Laurence. 2005. Conclusions on RGGI Allowances. Presentation to the RGGI Stakeholder Meeting, April 6. On file with author.

DeWitt, Laurence. 2011. Interview, February 11.

DeWitt, Laurence. 2013. Interview, October 25.

Dimock, Michael, Jocelyn Kiley, Scott Keeter, Carroll Doherty, and Alec Tyson. 2014. *Beyond Red vs. Blue: The Political Typology*. Washington, DC: Pew Research Center.

Dirix, Jo, Wouter Peeters, Johan Eyckmans, Peter Tom Jones, and Sigrid Sterckx. 2013. Strengthening Bottom-Up and Top-Down Climate Governance. *Climate Policy* 13 (3): 363–383.

Dominion. 2006. Public Comments on RGGI Draft Model Rule, May 19. http://www.rggi.org/design/history/public_comments (accessed November 28, 2015).

Doniger, David D. 1985. The Dark Side of the Bubble. *Environmental Forum* 4 (3): 33–35.

Druckman, James N. 2004. Political Preference Formation: Competition, Deliberation, and the (Ir)relevance of Framing Effects. *American Political Science Review* 98 (4): 671–686.

Dudek, Daniel J., and John Palmisano. 1988. Emissions Trading: Why Is This Thoroughbred Hobbled? *Columbia Journal of Environmental Law* 13:217–256.

"E&E's Power Plan Hub." 2015. *E&E News*. http://www.eenews.net/interactive/clean_power_plan (accessed December 5, 2015).

Eastman Kodak. 2006. Public Comments on RGGI Draft Model Rule, May 22. http://www.rggi.org/design/history/public_comments (accessed November 28, 2015).

Edison Electric Institute. 2005. SWG Proposal Comments, September 20. http://www.rggi.org/design/history/public_comments (accessed November 28, 2015).

Edison Electric Institute. 2006. Public Comments on RGGI Draft Model Rule, May 22. http://www.rggi.org/design/history/public_comments (accessed November 28, 2015).

"Editorial: Efficiency Required." 2007. *Bangor Daily News*, February 13.

"Editorial: How California Should Budget for Climate Change." 2014. *Los Angeles Times*, May 7.

Eheart, J. Wayland, E. Downey Brill, and Randolph M. Lyon. 1983. Transferable Discharge Permits for Control of BOD: An Overview. In *Buying a Better Environment: Cost-Effective Regulation through Permit Trading*, ed. Erhard F. Joeres and Martin H. David, 163–195. Madison: University of Wisconsin Press.

Ellerman, A. Denny. 2004. Allocation in the OTC NO_x Budget Program. Paper presented at the RGGI Stakeholder Workshop on Allocation, October 14. http://www.rggi.org/docs/ellerman_pres_10_14_04.pdf (accessed November 19, 2015).

Ellerman, A. Denny, Frank J. Convery, Christian De Perthuis, Emilie Alberola, Barbara K. Buchner, Anaïs Delbosc, Cate Hight, Jan Keppler, and Felix Matthes. 2010. *Pricing Carbon: The European Union Emissions Trading Scheme*. Cambridge: Cambridge University Press.

Ellerman, A. Denny, Paul L. Joskow, Richard Schmalensee, Juan-Pablo Montero, and Elizabeth M. Bailey. 2000. *Markets for Clean Air: The U.S. Acid Rain Program*. Cambridge: Cambridge University Press.

Ellickson, Robert C. 1991. *Order without Law: How Neighbors Settle Disputes*. Cambridge, MA: Harvard University Press.

Elster, Jon. 1989. *The Cement of Society: A Study of Social Order*. Cambridge: Cambridge University Press.

Elster, Jon. 1992. *Local Justice: How Institutions Allocate Scarce Goods and Necessary Burdens*. New York: Russell Sage Foundation.

Elster, Jon, ed. 1995. *Local Justice in America*. New York: Russell Sage Foundation.

Elster, Jon. 2007. *Explaining Social Behavior: More Nuts and Bolts for the Social Sciences*. Cambridge: Cambridge University Press.

Engel, Kirsten H., and Barak Y. Orbach. 2008. Reshaping the Global Warming Debate: Micro-Motives and State and Local Climate Change Initiatives. *Harvard Law and Policy Review* 2:119–137.

Entergy. 2006. Public Comments on RGGI Draft Model Rule, May 23. http://www.rggi.org/design/history/public_comments (accessed November 28, 2015).

Entman, Robert M. 1993. Framing: Towards Clarification of a Fractured Paradigm. *Journal of Communication* 43 (4): 51–58.

Environment Northeast. 2006a. Comments on RGGI Draft Model Rule, May 19. http://www.rggi.org/design/history/public_comments (accessed November 28, 2015).

Environment Northeast. 2006b. RGGI Consumer Benefit Allocation, August 1. On file with author.

Environment Northeast and Pace Energy Project. 2004. Public comment, Environment Northeast Draft RGGI Model Rule Outline, Key Issues, and Next Steps for Modeling, November 19. http://www.rggi.org/design/history/public_comments (accessed November 25, 2015).

Environment Northeast et al. 2004. Public Comment on RGGI Modeling Guidelines and Preliminary Comments, April 6. http://www.rggi.org/design/history/public_comments (accessed November 23, 2015).

Environment Northeast et al. 2005a. Public Comment, Incorporating Energy Efficiency Programs into RGGI Design, April 20. http://www.rggi.org/design/history/public_comments (accessed November 28, 2015).

Environment Northeast et al. 2005b. Public Comments on IPM Modeling Runs to Date, March 9. http://www.rggi.org/design/history/public_comments (accessed November 28, 2015).

Environment Northeast et al. 2005c. Public Comment, SWG Proposal Comments, September 12. http://www.rggi.org/design/history/public_comments (accessed November 28, 2015).

Environmental Advocates of New York. 2006. Comments on RGGI Draft Model Rule, May 22. http://www.rggi.org/design/history/public_comments (accessed November 28, 2015).

Environmental Advocates of New York et al. 2005. Public Comment, Model Rule, February 2. http://www.rggi.org/design/history/public_comments (accessed November 28, 2015).

Environmental Defense Fund. 2006. Comment on RGGI Draft Model Rule, May 23. http://www.rggi.org/design/history/public_comments (accessed November 23, 2015).

Environmental Defense Fund. 2012. Invest to Grow: Investing AB 32 Proceeds to Grow California's Clean and Efficient Economy. http://edf.org/invest-to-grow (accessed December 4, 2015).

Esarey, Justin, Timothy C. Salmon, and Charles Barrilleaux. 2012. What Motivates Political Preferences? Self-Interest, Ideology, and Fairness in a Laboratory Democracy. *Economic Inquiry* 50 (3): 604–624.

"EU Experience Seen Leading US to Auction, Not Give Away, Carbon Emission Allowances." 2007. *Electric Utility Week*, April 16.

European Commission. 1998. Climate Change—Towards an EU Post-Kyoto Strategy. Communication from the Commission to the Council and the European Parliament.COM (98) 353 final. http://aei.pitt.edu/6815/ (accessed November 20, 2015).

European Commission. 1999. Preparing for Implementation of the Kyoto Protocol. COM (99) 230. http://ec.europa.eu/clima/policies/ets/docs/com_1999_230_en.pdf (accessed November 20, 2015).

European Commission. 2000. Greenhouse Gas Emissions Trading and Climate Change Programme. http://eur-lex.europa.eu/legal-content/EN/TXT/HTML/?uri=UR ISERV:l28109&from=EN (accessed November 20, 2015).

European Commission. 2001a. Proposal for a Directive of the European Parliament and of the Council Establishing a Scheme for Greenhouse Gas Emission Allowance Trading Within the Community, COM(2001)581, October 23. http://eur-lex.europa .eu/legal-content/EN/TXT/HTML/?uri=CELEX:52001PC0581&from=EN.

European Commission. 2001b. Summary of Submissions: Green Paper on Greenhouse Gas Emissions Trading within the European Union, May 14. http://ec.europa. eu/environment/archives/docum/pdf/0087_summary.pdf (accessed November 20, 2015).

European Commission. 2014. Report to the European Parliament and the Council on Progress towards Achieving the Kyoto and EU 2020 Objectives. http://eur-lex. europa.eu/resource.html?uri=cellar:eb290b32-5e8e-11e4-9cbe-01aa75ed71a1 .0019.03/DOC_1&format=PDF (accessed December 1, 2015).

European Council. 2003. Communication from the Commission to the European Parliament. http://eur-lex.europa.eu/legal-content/EN/TXT/HTML/?uri=CELEX:520 03SC0364&from=EN (accessed November 22, 2015).

European Parliament. 2000. Report on the Commission Green Paper on Greenhouse Gas Emissions Trading within the European Union. October 11. http://ec.europa.eu/ environment/archives/docum/pdf/0087_governmental.pdf (accessed November 20, 2015).

European Parliament. 2002a. Debates, Thursday October 10, 2002—Brussels. http:// www.europarl.europa.eu/sides/getDoc.do?pubRef=-//EP//TEXT+CRE+20021010 +ITEM-001+DOC+XML+V0//EN&language=EN (accessed November 22, 2015).

European Parliament. 2002b. Report on the Proposal for a European Parliament and Council Directive Establishing a Scheme for Greenhouse Gas Emission Allowance

Trading within the Community and Amending Council Directive 96/61/EC, September 13. http://www.europarl.europa.eu/sides/getDoc.do?type=REPORT&mode= XML&reference=A5-2002-303&language=EN (accessed November 20, 2015).

Farrell, Alexander E. 2000. The NO$_x$ Budget: A Look at the First Year. *Electricity Journal* 13 (2): 83–93.

Farrell, Alexander E. 2001a. Multi-Lateral Emission Trading: Lessons from Inter-State NO$_x$ Control in the United States. *Energy Policy* 29 (13): 1061–1072.

Farrell, Alexander E. 2001b. The Political Economy of Interstate Public Policy: Power-Sector Restructuring and Transboundary Air Pollution. In *Improving Regulation: Cases in Environment, Health, and Safety*, ed. Paul S. Fischbeck and Scott Farrow, 115–141. Washington, DC: Resources for the Future Press.

Farrell, Alexander E., and W. Michael Hanemann. 2009. Field Notes on the Political Economy of California Climate Policy. In *Changing Climates in North American Politics: Institutions, Policymaking, and Multilevel Governance*, ed. Henrik Selin and Stacy D. VanDeveer, 87–109. Cambridge, MA: MIT Press.

Fehr, Ernst, and Urs Fischbacher. 2004. Social Norms and Human Cooperation. *Trends in Cognitive Sciences* 8 (4): 185–190.

Feldman, Stanley. 2003. Values, Ideology, and the Structure of Political Attitudes. In *Oxford Handbook of Political Psychology*, ed. David O. Sears, Leonie Huddy, and Robert Jervis, 477–508. Oxford: Oxford University Press.

Finnemore, Martha, and Kathryn Sikkink. 1998. International Norm Dynamics and Political Change. *International Organization* 52 (4): 887–917.

Fishbein, Martin, and Icek Azjen. 2010. *Predicting and Changing Human Behavior: The Reasoned Action Approach*. New York: Psychology Press.

Fleckenstein, Timo. 2011. The Politics of Ideas in Welfare State Transformation: Christian Democracy and the Reform of Family Policy in Germany. *Social Policy* 18 (4): 543–571.

Foundation for International Environmental Law and Development. 2000. Designing Options for Implementing an Emissions Trading Regime for Greenhouse Gases in the EC, February 22. http://ec.europa.eu/environment/archives/docum/ pdf/0087_field.pdf (accessed November 20, 2015).

Fowler, James H., and Nicholas A. Christakis. 2013. A Random World Is a Fair World. *Proceedings of the National Academy of Sciences of the United States of America* 110 (7): 2440–2441.

Franco-Watkins, Ana M., Bryan D. Edwards, and Roy E. Acuff. 2013. Effort and Fairness in Bargaining Games. *Journal of Behavioral Decision Making* 26 (1): 79–90.

Friedman, Elisabeth Jay, Kathryn Hochstetler, and Ann Marie Clark. 2005. *Sovereignty, Democracy, and Global Civil Society*. Albany: State University of New York Press.

Gabel, Steve. 2006. Descending Clock Auction for New Jersey BGS Power Procurements. Presentation to the RGGI Allowance Auction Workshop, July 20. http://www.rggi.org/design/history/topical_workshops (accessed November 28, 2015).

Gaines, Sanford E. 1991. The Polluter Pays Principle: From Economic Equity to Environmental Ethos. *Texas International Law Journal* 26:463–496.

Galizzi, Paolo, Dale Bryk, Jared Snyder, James Tripp, and Sean Donahue. 2006. Symposium: Reducing Greenhouse Gases: State Initiatives and Market-Based Solutions. *Fordham Environmental Law Review* 17:111–159.

Gallucci, Maria. 2012a. "California Torn Over How to Spend Cap-and-Trade Riches." *Inside Climate News*, June 12. http://insideclimatenews.org/news/20120612/california-cap-and-trade-program-billions-auction-proceeds-energy-efficiency-carb-jerry-brown-next10-rggi (accessed December 4, 2015).

Gallucci, Maria. 2012b. "Cap and Trade Resurrected? Some States Awaken to Its Economic Benefits." *Inside Climate News*, July 12. http://insideclimatenews.org/news/20120708/cap-and-trade-rgg-states-california-economic-benefits-energy-efficiency-jobs-carbon-auctions-proceeds-deficits (accessed December 1, 2015).

German NGO Forum for Environment and Development. 2000. German NGO Comments on EU-Green Paper on Greenhouse Gas Emissions Trading within the EU. September. http://ec.europa.eu/environment/archives/docum/pdf/0087_ngo.pdf (accessed November 20, 2015).

Gillard, Julia. 2012. Australia's Clean Energy Future. Press release, July 1. http://parlinfo.aph.gov.au/parlInfo/search/display/display.w3p;query=Id%3A%22media%2Fpressrel%2F1751751%22 (accessed December 5, 2015).

Glatt, Sandy. 2010. *Public Benefit Funds: Increasing Renewable Energy and Industrial Energy Efficiency Opportunities*. Golden, CO: US Department of Energy.

Gorman, Hugh S., and Barry D. Solomon. 2002. The Origins, Practice, and Limits of Emissions Trading. *Journal of Policy History* 14 (3): 293–320.

Greenwald, Judi. 2004. Allocation Choices: The U.S. Acid Rain Program. Presentation at RGGI Stakeholder Workshop on Allowance Apportionment and Allocation, October 14. http://www.rggi.org/docs/greenwald2_pres_10_14_04.pdf (accessed November 19, 2015).

Groppe, Maureen. 2015. "Pence: EPA Must Change Emissions Rules or Indiana Won't Comply." *Indianapolis Star*, June 24. http://www.indystar.com/story/news/politics/2015/06/24/pence-epa-emissions-rules-need-changes-indiana-agree/29215723/ (accessed December 5, 2015).

Grubb, Michael. 2012. Cap and Trade Finds New Energy. *Nature* 491:666–667.

Hahn, Robert W. 1983. Designing Markets in Transferable Property Rights: A Practitioner's Guide. In *Buying a Better Environment: Cost-Effective Regulation through Permit Trading*, ed. Erhard F. Joeres and Martin H. David, 83–97. Madison: University of Wisconsin Press.

Hahn, Robert W. 1989. Economic Prescriptions for Environmental Problems: How the Patient Followed the Doctor's Orders. *Journal of Economic Perspectives* 3:95–114.

Hamel, Sonia. 2010. Interview, November 5.

Hamel, Sonia. 2013. Interview, October 3.

Hannesson, Rögnvaldur. 2004. *The Privatization of the Oceans*. Cambridge, MA: MIT Press.

Harrison, David, and Daniel B. Radov. 2002. Evaluation of Alternative Initial Allocation Mechanisms in a European Union Greenhouse Gas Emissions Allowance Trading Scheme. National Economic Research Associates. http://www.merlin-project.de/restricted/sr_workspace/EU%20Emission%20Trading%20evaluation.pdf (accessed November 20, 2015).

Harrison, Kathryn. 2010. The Comparative Politics of Carbon Taxation. *Annual Review of Law and Social Science* 6:507–529.

Harrison, Kathryn. 2012. A Tale of Two Taxes: The Fate of Environmental Tax Reform in Canada. *Review of Policy Research* 29 (3): 383–407.

Hartz, Louis. 1955. *The Liberal Tradition in America*. New York: Harcourt, Brace.

Harvey, Fiona. 2013. "Emissions Trading Scheme: EU Committee Passes 'Rescue' Reforms." *Guardian*, February 19. http://www.theguardian.com/environment/2013/feb/19/emissions-trading-scheme-eu-rescue-reforms (accessed December 1, 2015).

Hazlett, Thomas W. 1998. Assigning Property Rights to Radio Spectrum Users: Why Did FCC License Auctions Take 67 Years? *Journal of Law and Economics* 41 (S2): 529–576.

Hechter, Michael, and Karl-Dieter Opp, eds. 2001. *Social Norms*. New York: Russell Sage Foundation.

Hegelich, Simon, Cornelia Fraune, and David Knollmann. 2015. Point Predictions and the Punctuated Equilibrium Theory: A Data Mining Approach—U.S. Nuclear Policy as Proof of Concept. *Policy Studies Journal* 43 (2): 228–256.

Heinmiller, Timothy. 2007. The Politics of "Cap and Trade" Policies. *Natural Resources Journal* 47:445–467.

Henrich, Joseph, Robert Boyd, Samuel Bowles, Colin Camerer, Ernst Fehr, and Herbert Gintis. 2004. *Foundations of Human Sociality: Economic Experiments and Ethnographic Evidence from Fifteen Small-Scale Societies.* Oxford: Oxford University Press.

Henrich, Joseph, Robert Boyd, Samuel Bowles, Colin Camerer, Ernst Fehr, Herbert Gintis, Richard McElreath, et al. 2005. "Economic Man" in Cross-Cultural Perspective: Behavioral Experiments in Fifteen Small-Scale Societies. *Behavioral and Brain Sciences* 28 (6): 795–815.

Hibbard, Paul J., Nancy L. Seidman, Barbara Finemore, and David Moskovitz. 2000. Output-Based Emission Control Programs: U.S. Experience. China Sustainable Energy Program. http://www.efchina.org/Attachments/Report/reports-efchina-2003 0509-1-en/EmitUSXp.pdf (accessed November 20, 2015).

Hibbard, Paul J., Susan F. Tierney, Andrea M. Okie, and Pavel G. Darling. 2011. The Economic Impact of the Regional Greenhouse Gas Initiative on Ten Northeast and Mid-Atlantic States. Analysis Group. http://www.analysisgroup.com/uploadedfiles/ content/insights/publishing/economic_impact_rggi_report.pdf (accessed December 1, 2015).

Hochschild, Jennifer L. 1981. *What's Fair? American Beliefs about Distributive Justice.* Cambridge, MA: Harvard University Press.

Holden, Emily. 2014. "RGGI May Not Be an Easy Model for Regional Groups to Comply with EPA's Carbon Rule." *ClimateWire*, November 25. http://www.eenews .net/stories/1060009523 (accessed December 1, 2015).

Holden, Emily, Rod Kuckro, and Peter Behr. 2015. "Clean Power Plan Changes Appease Many Concerns, But Coal Lobby Promises a Fight." *EnergyWire*, August 3. http://www.eenews.net/stories/1060022870 (accessed December 5, 2015).

Holt, Charles. 2006. Auctions and Auctioneering: Public Policy Applications. Presentation at the RGGI Allowance Auction Workshop, July 20. http://www.rggi.org/ design/history/topical_workshops (accessed November 19, 2015).

Hope, Mat. 2014. "Why a Healthy Carbon Market Underpins the EU's 2030 Climate Goals." *Carbon Brief*, October 21. http://www.carbonbrief.org/blog/2014/10/why -unofficial-carbon-market-discussions-are-key-to-eu-climate-negotiations/ (accessed December 1, 2015).

Htun, Mala, and S. Laurel Weldon. 2012. The Civic Origins of Progressive Policy Change: Combating Violence against Women in Global Perspective. *American Political Science Review* 106:548–569.

Huber, Bruce R. 2013. How Did RGGI Do It? Political Economy and Emissions Auctions. *Ecology Law Quarterly* 40:59–106.

"Impact of Pollution Plan Debated; Industry Warns of Higher Bills If State Adopts Credit System, but Supporters Say It's Fair." 2006. *Times Union* (Albany, NY), December 13.

Independent Energy Producers of New Jersey. 2006. Public Comments on RGGI Draft Model Rule, May 18. http://www.rggi.org/design/history/public_comments (accessed November 28, 2015).

Independent Power Producers of New York. 2006. Public Comments on RGGI Draft Model Rule, May 22. http://www.rggi.org/design/history/public_comments (accessed November 28, 2015).

International Carbon Action Partnership. 2015. ETS Fact Sheets, July 20. https://icapcarbonaction.com/ets-map (accessed December 1, 2015).

Jacoby, William G. 2000. Issue Framing and Public Opinion on Government Spending. *American Journal of Political Science* 44 (4): 750–767.

Jenkins-Smith, Hank C., Daniel Nohrstedt, Christopher M. Weible, and Paul A. Sabatier. 2014. The Advocacy Coalition Framework: Foundations, Evolution, and Ongoing Research. In *Theories of the Policy Process*, ed. Paul A. Sabatier and Christopher M. Weible, 183–223. Boulder, CO: Westview Press.

Jenkins-Smith, Hank C., Gilbert K. St. Clair, and Brian Woods. 1991. Explaining Change in Policy Subsystems: Analysis of Coalitions Stability and Defection over Time. *American Journal of Political Science* 35 (4): 851–880.

Joint Environmental Organizations. 2005. Letter, SWG Proposal Comments, September 13. http://www.rggi.org/design/history/public_comments (accessed November 28, 2015).

Joint Utility Consumer Advocates. 2005. Public Comment, Recommendations (ME, CT, NH), April 19. http://www.rggi.org/design/history/public_comments (accessed November 28, 2015).

Jones, Bryan D., and Frank R. Baumgartner. 2005. *The Politics of Attention*. Chicago: University of Chicago Press.

Jones, Bryan D., and Frank R. Baumgartner. 2012. From There to Here: Punctuated Equilibrium to the General Punctuation Thesis to a Theory of Government Information Processing. *Policy Studies Journal* 40 (1): 1–19.

Joskow, Paul L., and Richard Schmalensee. 1998. The Political Economy of Market-Based Environmental Policy: The U.S. Acid Rain Program. *Journal of Law and Economics* 41:37–83.

Kahan, Dan. 2010. Fixing the Communications Failure. *Nature* 463 (7279): 296–297.

Kahan, Dan M. 2015. Climate-Science Communication and the Measurement Problem. *Political Psychology* 36 (S1): 1–43.

Kahn, Debra. 2015. "Calif. Engaging in Carbon-Trading Discussions with Other Western States." *ClimateWire*, August 21. http://www.eenews.net/stories/1060023751 (accessed December 5, 2015).

Kantner, James. 2011. "E.U. Closes Emissions Trading System after Thefts." *New York Times*, January 19. http://www.nytimes.com/2011/01/20/business/global/20iht -carbon20.html (accessed December 1, 2015).

Kaplan, Seth. 2011. Interview, April 20.

Kaplan, Seth. 2014. Interview, June 9.

Kaye, Leon. 2014. California Readies for Cap-and-Trade Next Steps. TriplePundit (blog), November 5. http://www.triplepundit.com/2014/11/cap-trade-california -readies-next-steps/ (accessed December 4, 2015).

Keohane, Nathaniel O., Richard L. Revesz, and Robert N. Stavins. 1998. The Choice of Regulatory Instruments in Environmental Policy. *Harvard Environmental Law Review* 22 (313–367): 313–367.

Kete, Nancy. 1992. The U.S. Acid Rain Control Allowance Trading System. In *Climate Change: Designing a Tradable Permit System*, ed. Organization for Economic Cooperation and Development. Paris: Organization for Economic Cooperation and Development.

Keyspan. 2006. Public Comments on RGGI Draft Model Rule, May 18. http://www .rggi.org/design/history/public_comments (accessed November 28, 2015).

Kharbanda, Jesse. 2015. Interview, August 26.

Kingdon, John W. 2003. *Agendas, Alternatives, and Public Policies*. 2nd ed. New York: Longman.

Klein, Naomi. 2014. *This Changes Everything: Capitalism vs. the Climate*. New York: Simon and Schuster.

Klinsky, Sonja. 2013. Bottom-up Policy Lessons Emerging from the Western Climate Initiative's Development Challenges. *Climate Policy* 13 (2): 143–169.

Klüver, Heike, and Christine Mahoney. 2015. Measuring Interest Group Framing Strategies in Public Policy Debates. *Journal of Public Policy* 35 (2): 223–244.

Kneese, Allen V., and Charles L. Schultze. 1975. *Pollution, Prices, and Public Policy*. Washington, DC: Brookings Institute.

Krolewski, Mary Jo, and Andrew S. Mingst. 2000. Recent NO_x Reduction Efforts: An Overview. Paper presented at the ICAC Forum, Arlington, Virginia, March.

Krosnick, Jon A., and Bo MacInnis. 2011. *National Survey of American Public Opinion on Global Warming*. Stanford, CA: Stanford University.

Kruger, Joe. 2004. Comparative Cap-and-Trade Programs: U.S. SO_2 and NO_x, EU CO_2. Presentation at the RGGI Stakeholder Meeting, April 2. On file with author.

Kushler, Martin. 1998. *An Updated Status Report of Public Benefit Programs in an Evolving Electric Utility Industry*. Washington, DC: American Council for an Energy Efficient Economy.

Kwasnik, Joseph. 2011. Interview, September 20.

Kwerel, Evan. 2006. Spectrum Auctions at the FCC. Presentation at the RGGI Allowance Auction Workshop, July 20. http://www.rggi.org/design/history/topical _workshops (accessed November 19, 2015).

Lachapelle, Erick, Christopher Borick, and Barry Rabe. 2012. Public Attitudes toward Climate Science and Climate Policy in Federal Systems: Canada and the United States Compared. *Review of Policy Research* 29 (3): 334–357.

Lakoff, George. 2002. *Moral Politics: How Liberals and Conservatives Think*. 2nd ed. Chicago: University of Chicago Press.

Lakoff, George. 2008. *The Political Mind: Why You Can't Understand Twenty-First-Century Politics with an Eighteenth-Century Brain*. New York: Viking.

Lamkin, William. 2012. Interview, October 3.

Lange, Andreas, Carsten Vogt, and Andreas Ziegler. 2007. On the Importance of Equity in International Climate Policy: An Empirical Analysis. *Energy Economics* 29 (3): 545–562.

Legro, Jeffrey W. 2000. The Transformation of Policy Ideas. *American Journal of Political Science* 44:419–432.

Leiserowitz, Anthony, Edward Maibach, Connie Roser-Renouf, Geoff Feinberg, Seth Rosenthal, and Jennifer Marlon. 2014. *Climate Change in the American Mind: October 2014*. New Haven, CT: Yale University Press.

Levenson, Marilyn. 2014. Personal communication, January 6.

Libecap, Gary D. 1989. *Contracting for Property Rights*. Cambridge: Cambridge University Press.

Lifsher, Mark. 2014. "Democratic Bill Would Slow California's Effort to Curb Climate Change." *Los Angeles Times*, July 13. http://www.latimes.com/business/la -fi-capitol-business-beat-20140714-story.html (accessed December 4, 2015).

Lim, Nathan. 2014. Did Anyone Notice That Australia Now Has a Carbon Trading Scheme? RenewEconomy (blog), December 17. http://reneweconomy.com.au/2014/

anyone-notice-australia-now-carbon-trading-scheme-87832 (accessed December 5, 2015).

Lindblom, Charles E. 1959. The Science of "Muddling Through." *Public Administration Review* 19 (2): 79–88.

Litz, Franz T. 2003. Personal Notes, SWG Meeting, September 11. On file with author.

Litz, Franz T. 2004a. Personal notes, SWG Meeting, January 15–16. On file with author.

Litz, Franz T. 2004b. Personal notes, SWG Meeting, October 18–19, Newport, RI. On file with author.

Litz, Franz T. 2011. Interview, March 22.

Litz, Franz T. 2013. Interview, October 2.

Litz, Franz T., and Jennifer Macedonia. 2015. Choosing a Policy Pathway for State 111(d) Plans to Meet State Objectives. *Great Plains Institute.* http://www.betterenergy.org/sites/www.betterenergy.org/files/Policy%20Pathways%20Paper.pdf.

Locke, John. (1690) 1994. *Two Treatises of Government.* Cambridge: Cambridge University Press.

Lockwood, Matthew. 2011. Does the Framing of Climate Policies Make a Difference to Public Support? Evidence from UK Marginal Constituencies. *Climate Policy* 11:1097–1112.

London, Jonathan, Alex Karner, Julie Sze, Dana Rowan, Gerardo Gambirazzio, and Deb Niemeier. 2013. Racing Climate Change: Collaboration and Conflict in California's Global Climate Change Policy Arena. *Global Environmental Change* 23 (4): 791–799.

Lord, Peter B. 2008. "Carbon Auction Nets $16 Million for Maryland." *Capital* (Annapolis, MD), September 30.

Loris, Nicolas. 2015. Four Big Problems with the Obama Administration's Climate Change Regulations. Issue Brief #4454 on Energy and Environment, August 14. http://www.heritage.org/research/reports/2015/08/four-big-problems-with-the-obama-administrations-climate-change-regulations (accessed December 5, 2015).

Lowi, Theodore J. 1979. *The End of Liberalism.* 2nd ed. New York: W. W. Norton and Company.

Lynch, Julia, and Sarah E. Gollust. 2010. Playing Fair: Fairness Beliefs and Health Policy Preferences in the United States. *Journal of Health Politics, Policy, and Law* 35 (6): 849–887.

MA Climate Action Network. 2006. Public Comment on RGGI Draft Model Rule, May 22. http://www.rggi.org/design/history/public_comments (accessed November 28, 2015).

MA Climate Coalition. 2004. Public Comment, June. http://www.rggi.org/design/history/public_comments (accessed November 23, 2015).

MA Department of Environmental Protection (MA DEP). 1999a. Background Document and Technical Support for Public Hearings on the Proposed Revisions to the State Implementation Plan for Ozone: Response to the "NOₓ SIP Call" and the "OTC NOₓ MOU," including Amendments to 310 CMR 7.00 et. seq.: 310 CMR 7.19 "RACT for Sources of Oxides of Nitrogen," 310 CMR 7.27, "NOₓ Allowance Program," and 310 CMR 7.28, "NOₓ Allowance Trading Program," July. On file with author.

MA Department of Environmental Protection (MA DEP). 1999b. Summary of Comments and Response to Comments from Public Hearings on Proposed Revisions to the State Implementation Plan for Ozone, including Proposed 310 CMR 7.28, November. On file with author.

MA Division of Energy Resources. 2007. Massachusetts Saving Electricity: A Summary of the Performance of Electric Efficiency Programs Funded by Ratepayers between 2003 and 2005, April 2. http://www.mass.gov/eea/docs/doer/electric-deregulation/ee03-05.pdf (accessed November 19, 2015).

Macinko, Seth. 2010. The View from Fisheries Management: Lessons Learned from "Catch Shares." Presentation at the Purdue Climate Change Research Center Emissions Trading Workshop, April 30. http://www.purdue.edu/discoverypark/climate/docs/Macinko.pdf (accessed January 12, 2016).

Macinko, Seth, and Daniel W. Bromley. 2002. *Who Owns America's Fisheries*. Washington, DC: Island Press.

Macinko, Seth, and Leigh Raymond. 2001. Fish on the Range: The Perils of Crossing Conceptual Boundaries in Natural Resource Policy. *Marine Policy* 25 (2): 123–131.

Maine Public Advocate. 2004. Public Comment, Macroeconomic Modeling, August 18. http://www.rggi.org/design/history/public_comments (accessed November 25, 2015).

Majone, Giandomenico. 1989. *Evidence, Argument, and Persuasion in the Policy Process*. New Haven, CT: Yale University Press.

Major, Mary. 2013. Interview, August 13.

Marietta, Morgan. 2008. From My Cold, Dead Hands: Democratic Consequences of Sacred Rhetoric. *Journal of Politics* 70 (3): 767–779.

Marshall, Christa. 2010. "Money to Fight Climate Change Gets Siphoned into Other Budgets." *ClimateWire*, March 19. http://www.eenews.net/climatewire/2010/03/19/stories/88904 (accessed December 1, 2015).

Massachusetts Launches CO_2 Control Plan with Price Caps, Separately from RGGI. 2005. *Electric Utility Week*, December 12.

Matisoff, Daniel C. 2010. Making Cap-and-Trade Work: Lessons from the European Experience. *Environment* 52 (1): 10–19.

May, Peter J. 2005. Regulation and Compliance Motivations: Examining Different Approaches. *Public Administration Review* 65 (1): 31–44.

Mays, Jon. 2014. "Cost of Cap-and-Trade." *San Mateo Daily Journal*, June 20. http://www.smdailyjournal.com/articles/opinions/2014-06-20/cost-of-cap-and-trade/1776425125264.html (accessed December 4, 2015).

McCall, Leslie. 2013. *The Underserving Rich: American Beliefs about Inequality, Opportunity, and Redistribution*. Cambridge: Cambridge University Press.

McCarthy, Gina. 2014. Remarks Announcing Clean Power Plan, June 2. http://yosemite.epa.gov/opa/admpress.nsf/8d49f7ad4bbcf4ef852573590040b7f6/c45baade030b640785257ceb003f3ac3!OpenDocument (accessed December 5, 2015).

McConnell, Grant. 1966. *Private Power and American Democracy*. New York: Alfred A. Knopf.

McCright, Aaron M., and Riley E. Dunlap. 2011. Cool Dudes: The Denial of Climate Change among Conservative White Males in the United States. *Global Environmental Change* 21 (4): 1163–1172.

McLean, Brian J. 1997. Evolution of Marketable Permits: The U.S. Experience with Sulfur Dioxide Allowance Trading. *International Journal of Environment and Pollution* 8:19–36.

Meckling, Jonas. 2011. *Carbon Coalitions: Business, Climate Politics, and the Rise of Emissions Trading*. Cambridge, MA: MIT Press.

Meier, Kenneth J. 2009. Policy Theory, Policy Theory Everywhere: Ravings of a Deranged Policy Scholar. *Policy Studies Journal* 37 (1): 5–11.

Menkes, Neal. 2014. Interview, August 13.

Mettler, Suzanne, and Joe Soss. 2004. The Consequences of Public Policy for Democratic Citizenship: Bridging Policy Studies and Mass Politics. *Perspectives on Politics* 2 (1): 55–73.

Midwestern Greenhouse Gas Reduction Accord (MGGRA). 2009. Advisory Group Draft Final Recommendations, June. On file with author.

Mikhail, John. 2011. *Elements of Moral Cognition: Rawls' Linguistic Analogy and the Cognitive Science of Moral and Legal Judgment*. Cambridge: Cambridge University Press.

Miller, Kevin. 2005. "Maine Joins Greenhouse Compact." *Bangor Daily News*, December 21.

Miller, Kevin. 2007. "Brits Share Advice on Greenhouse Gas; Parliament Officials Meet with Baldacci." *Bangor Daily News*, March 14.

Montgomery, W. David. 1972. Markets in Licenses and Efficient Pollution Control Programs. *Journal of Economic Theory* 5:395–418.

Morain, Dan. 2014. "Lining Up for the Cap and Trade Gusher." *Sacramento Bee*, March 23. http://www.sacbee.com/opinion/opn-columns-blogs/dan-morain/article 2593583.html (accessed December 4, 2015).

Moss, David A., and Michael R. Fein. 2003. Radio Regulation Revisited: Coase, the FCC, and the Public Interest. *Journal of Policy History* 15 (4): 389–416.

Mucciaroni, Gary. 2011. Are Debates about "Morality Policy" Really about Morality? Framing Opposition to Gay and Lesbian Rights. *Policy Studies Journal* 39 (2): 187–261.

Mulkern, Anne. 2014. "Gov. Brown and Lawmakers Reach Deal on Spending Billions in Carbon Cap-and-Trade Revenues." *ClimateWire*, June 13. http://www .eenews.net/stories/1060001261 (accessed December 4, 2015).

Multiple Environmental NGOs. 2006. Comment on Draft RGGI Model Rule, April 20. http://www.rggi.org/design/history/public_comments (accessed November 28, 2015).

Multiple Intervenors. 2006. Comments on RGGI Draft Model Rule, May 23. http:// www.rggi.org/design/history/public_comments (accessed November 28, 2015).

Murray, William. 2001. Memo to Governor Mark Warner, May 17. On file with author.

Murray, William. 2013. Interview, August 29.

Murrow, Derek. 2010. Interview, December 1.

Murrow, Derek. 2013. Interview, November 6.

Napolitano, Sam, Gabrielle Stevens, Jeremy Schreifels, and Kevin Culligan. 2007. The NO_x Budget Trading Program: A Collaborative, Innovative Approach to Solving a Regional Air Pollution Problem. *Electricity Journal* 20 (9): 65–76.

Nash, Jonathan R. 2000. Too Much Market? Conflict between Tradable Pollution Allowances and the "Polluter Pays" Principle. *Harvard Environmental Law Review* 24:465–535.

National Association of State PIRGs. 2005a. Offsets Comments, September 20. http://www.rggi.org/design/history/public_comments (accessed November 28, 2015).

National Association of State PIRGs. 2005b. Public Comments, Role of Energy Efficiency in Reducing CO$_2$ Emissions in the NE, August 24. http://www.rggi.org/design/history/public_comments (accessed November 28, 2015).

National Grid. 2005a. Public Comment, Allowance Allocation, August 29. http://www.rggi.org/design/history/public_comments (accessed November 28, 2015).

National Grid. 2005b. Public Comment, Allowance Allocation White Paper, September 20. http://www.rggi.org/design/history/public_comments (accessed November 28, 2015).

National Grid. 2006. Public Comments on RGGI Draft Model Rule, May 22. http://www.rggi.org/design/history/public_comments (accessed November 28, 2015).

Natural Resources Defense Council (NRDC). 2006. Public Comment on RGGI Draft Model Rule, May 22. http://www.rggi.org/design/history/public_comments (accessed November 28, 2015).

Nature Conservancy. 2006. Public Comment on RGGI Draft Model Rule, May 22. http://www.rggi.org/design/history/public_comments (accessed November 28, 2015).

Navarro-Treichler, Lauren. 2014. How California Produced the State's First Ever Climate Dividend. EDF Voices (blog), April 16. http://www.edf.org/blog/2014/04/16/how-california-produced-states-first-ever-climate-dividend (accessed December 4, 2015).

Nearing, Brian. 2006. "Division over Payments for Pollution Credits; Power Plants Want Them for Free, but Consumers Object." *Times Union* (Albany, NY), December 11.

Nedelsky, Jennifer. 1990. *Private Property and the Limits of American Constitutionalism.* Chicago: University of Chicago Press.

Nelson, Thomas E., Rosalee A. Clawson, and Zoe M. Oxley. 1997. Media Framing of a Civil Liberties Conflict and Its Effect on Tolerance. *American Political Science Review* 91 (3): 567–583.

Nelson, Thomas E., and Donald R. Kinder. 1996. Issue Frames and Group-Centrism in American Public Opinion. *Journal of Politics* 58 (4): 1055–1078.

Nelson, Thomas E., and Zoe M. Oxley. 1999. Issue Framing Effects on Belief Importance and Opinion. *Journal of Politics* 61 (4): 1040–1067.

Nelson, Thomas E., Zoe M. Oxley, and Rosalee A. Clawson. 1997. Toward a Psychology of Framing Effects. *Political Behavior* 19 (3): 221–246.

Neslen, Arthur. 2014. "EU Leaders Agree to Cut Greenhouse Gas Emissions by 40% by 2030." *Guardian*, October 23. http://www.theguardian.com/world/2014/oct/24/

eu-leaders-agree-to-cut-greenhouse-gas-emissions-by-40-by-2030 (accessed December 1, 2015).

New England Business Council et al. 2006. Public Comments on RGGI Draft Model Rule, May 23. http://www.rggi.org/design/history/public_comments (accessed November 28, 2015).

New England Governors/Eastern Canadian Premiers (NEG/ECP). 2001. Climate Change Action Plan, August. On file with author.

"New Jersey Legislature Passes RGGI Bill Requiring 100% Auctions for Allowances." 2008. *Electric Utility Week*, January 14.

"New York 100% Auction Plan for CO₂ Rights Meets IPPNY Resistance, but State Defends It." 2006. *Electric Utility Week*, December 11.

Newell, Richard G., William A. Pizer, and Daniel Raimi. 2013. Carbon Markets 15 Years after Kyoto: Lessons Learned, New Challenges. *Journal of Economic Perspectives* 27 (1): 123–146.

Nichols, Albert L. 1996. Designing a Trading Programme for Emissions of Nitrogen Oxides in the Northeastern United States. In *Environmental Policy between Regulation and Market*, ed. Claude Jeanrenaud, 171–197. Basel: Birkhauser Verlag.

Nisbet, Matthew C. 2009. Communicating Climate Change: Why Frames Matter for Public Engagement. *Environment* 51 (2): 12–23.

Nisbet, Matthew C., and Teresa Myers. 2007. Trends: Twenty Years of Public Opinion about Global Warming. *Public Opinion Quarterly* 71 (3): 444–470.

Northeast Regional GHG Coalition. 2004a. Public Comment, Allowance Allocation Methodologies, October 11. http://www.rggi.org/design/history/public_comments (accessed November 25, 2015).

Northeast Regional GHG Coalition. 2004b. Public Comments—Emission Portfolio Standards, March 12. http://www.rggi.org/design/history/public_comments (accessed November 25, 2015).

Northeast Regional GHG Coalition. 2004c. Public Comment—Key RGGI Principles, December 13. http://www.rggi.org/design/history/public_comments (accessed November 25, 2015).

Northeast Regional GHG Coalition. 2006. Public Comments on RGGI Draft Model Rule, May 22. http://www.rggi.org/design/history/public_comments (accessed November 28, 2015).

Northeast Suppliers. 2006. Public Comments on RGGI Draft Model Rule, May 23. http://www.rggi.org/design/history/public_comments (accessed November 28, 2015).

Novak, Robert. 2006. "Romney Gains by Shunning CO_2 Caps." *Chicago Sun Times,* January 2.

Nowak, Martin A., Karen M. Page, and Karl Sigmund. 2000. Fairness versus Reason in the Ultimatum Game. *Science* 289 (5485): 1773–1775.

NRG Energy. 2006. Public Comments on RGGI Draft Model Rule, May 22. http://www.rggi.org/design/history/public_comments (accessed November 28, 2015).

Nuccitelli, Dana. 2014. "The Latest Global Warming Bill and the Republican Conundrum." *Guardian,* November 25. http://www.theguardian.com/environment/climate-consensus-97-per-cent/2014/nov/25/latest-global-warming-bill-republican-conundrum (accessed December 6, 2015).

Nussbaum, Alex. 2008. "Plan to Reduce Greenhouse Gas Called a Sellout; Panels Pass Bills Allowing Utilities to Raise Their Rates." *The Record* (Bergen County, NJ), January 4.

NY Attorney General's Office. 2006. Comments on RGGI Draft Model Rule, May 23. http://www.rggi.org/design/history/public_comments (accessed November 28, 2015).

NY Coalition. 2004. Detailed Comments, April 2. http://www.rggi.org/design/history/public_comments (accessed December 6, 2015).

NY Coalition of Energy and Business Groups. 2006. Public Comments on RGGI Draft Model Rule, May 22. http://www.rggi.org/design/history/public_comments (accessed November 28, 2015).

NY Energy Consumers Council. 2006. Comments on RGGI Draft Model Rule, May 22. http://www.rggi.org/design/history/public_comments.

NY Environmental Coalition. 2004. Public Comment on RGGI Platform, May 18. http://www.rggi.org/design/history/public_comments (accessed November 23, 2015).

NY Public Service Commission. 1996. Opinion and Order regarding Competitive Opportunities for Electric Service, Opinion No. 96–12, Case 94-E-0952, May 20. On file with author.

NYC Economic Development Corporation. 2006. Public Comment on RGGI Draft Model Rule, May 22. http://www.rggi.org/design/history/public_comments (accessed November 28, 2015).

Obama, Barack. 2015. Remarks by the President in Announcing the Clean Power Plan, August 3. https://www.whitehouse.gov/the-press-office/2015/08/03/remarks-president-announcing-clean-power-plan (accessed December 5, 2015).

Olive, Andrea. 2014. *Land, Stewardship, and Legitimacy: Endangered Species Policy in Canada and the United States.* Toronto: University of Toronto Press.

Olson, Mancur. 1965. *The Logic of Collective Action*. Cambridge, MA: Harvard University Press.

Onuf, Nicholas. 1998. Constructivism: A User's Manual. In *International Relations in a Constructed World*, ed. Vendulka Kualkova, Nicholas Onuf and Paul Kowart. Armonk, NY: M. E. Sharpe.

Ostrom, Elinor. 1990. *Governing the Commons: The Evolution of Institutions for Collective Action*. Cambridge: Cambridge University Press.

Ostrom, Elinor. 1998. A Behavioral Approach to the Rational Choice Theory of Collective Action: Presidential Address, American Political Science Association, 1997. *American Political Science Review* 92 (1): 1–22.

"Other States Seen Taking New York Approach on RGGI Allowances, to the Market's Chagrin." 2006. *Electric Utility Week*, December 25.

Pace Energy Law Project. 2006. Comments on RGGI Draft Model Rule, May 22. http://www.rggi.org/design/history/public_comments (accessed November 28, 2015).

Pacheco, Marc. 2014. Interview, July 18.

Palmer, Karen. 2006. Introduction to Workshop to Support the Regional Greenhouse Gas Initiative on the Topic: Implementing the Minimum 25% Public Benefit Allocation. Presentation to the RGGI Allowance Auction Workshop, July 20. http://www.rggi.org/design/history/topical_workshops (accessed November 28, 2015).

Palmer, Karen. 2010. Allowance Allocation and Effects on the Electricity Sector. Presentation to the Purdue Climate Change Research Center Emissions Trading Workshop, April 30. http://www.purdue.edu/discoverypark/climate/docs/Palmer.pdf (accessed January 18, 2016).

Pataki, George. 2014. Interview, April 21.

"Patrick Signs Up Bay State for Battle vs. Global Warming." 2007. *Lowell Sun*, January 21.

Paylor, David. 2013. Interview, September 20.

Pearce, David. 2002. An Intellectual History of Environmental Economics. *Annual Review of Energy and the Environment* 27 (1): 57–81.

Pearlstein, Steven. 2009. "Climate-Change Bill Hits Some of the Right Notes But Botches the Refrain." *Washington Post*, May 22. http://www.washingtonpost.com/wp-dyn/content/article/2009/05/21/AR2009052104402.html (accessed December 4, 2015).

Pérez Henríquez, Blas Luis. 2013. *Environmental Commodities Markets and Emissions Trading: Toward a Low-Carbon Future*. New York: Resources for the Future Press.

Peters, B. Guy, Jon Pierre, and Desmond S. King. 2005. The Politics of Path Dependence: Political Conflict in Historical Institutionalism. *Journal of Politics* 67 (4): 1275–1300.

Petty, Richard E., and John T. Cacioppo. 1986. The Elaboration Likelihood Model of Persuasion. *Advances in Experimental Social Psychology* 19:123–162.

Pew Center on Global Climate Change. 2008. Distribution of Allowances, S. 3036 Boxer-Lieberman-Warner Substitute Amendment, June 3. http://www.c2es.org/docUploads/S3036-allocation-summary.pdf (accessed December 4, 2015).

Pew Center on Global Climate Change. 2009. Distribution of Allowances under the American Clean Energy and Security Act (Waxman-Markey), August. http://www.c2es.org/docUploads/policy-memo-allowance-distribution-under-waxman-markey.pdf (accessed December 4, 2015).

Pierson, Paul. 2000. Increasing Returns, Path Dependence, and the Study of Politics. *American Political Science Review* 94 (2): 251–267.

Pigou, Arthur C. 1920. *The Economics of Welfare*. London: Macmillan.

Povich, Elaine S. 2015. Plummeting Oil Price Creates Problems for Severance-Tax States. Stateline blog, Pew Charitable Trusts, February 5. http://www.pewtrusts.org/en/research-and-analysis/blogs/stateline/2015/2/05/plummeting-oil-price-creates-problems-for-severance-tax-states (accessed November 19, 2015).

Pralle, Sarah B. 2006. *Branching Out, Digging In: Environmental Advocacy and Agenda Setting*. Washington, DC: Georgetown University Press.

Public Service Enterprise Group. 2006. Public Comments on RGGI Draft Model Rule, May 22. http://www.rggi.org/design/history/public_comments (accessed November 28, 2015).

Rabe, Barry G. 2004. *Statehouse and Greenhouse*. Washington, DC: Brookings Institute.

Rabe, Barry G. 2008a. Regionalism and Global Climate Change Policy. In *Intergovernmental Management for the Twenty-First Century*, ed. Timothy J. Conlan and Paul L. Posner, 176–205. Washington, DC: Brookings.

Rabe, Barry G. 2008b. States on Steroids: The Intergovernmental Odyssey of American Climate Policy. *Review of Policy Research* 25 (2): 105–128.

Rabe, Barry G. 2010. The Aversion to Direct Cost Imposition: Selecting Climate Policy Tools in the United States. *Governance: An International Journal of Policy, Administration, and Institutions* 23 (4): 583–608.

Rabe, Barry G. 2013. Building on Sub-Federal Climate Strategies: The Challenges of Regionalism. In *Climate Change Policy in North America: Designing Integration in a*

Regional System, ed. A. Neil Craik, Isabel Studer, and Debora Van Nijnatten, 71–107. Toronto: University of Toronto Press.

Rabe, Barry G. 2014. Shale Play Politics: The Intergovernmental Odyssey of American Shale Governance. *Environmental Science and Technology* 48 (15): 8369–8375.

Rabe, Barry G. 2015. The Durability of Carbon Cap-and-Trade Policy. *Governance: An International Journal of Policy, Administration, and Institutions*. http://onlinelibrary .wiley.com/doi/10.1111/gove.12151/epdf (accessed December 4, 2015).

Rabe, Barry G., and Christopher P. Borick. 2012. Carbon Taxation and Policy Labeling: Experience from American States and Canadian Provinces. *Review of Policy Research* 29 (3): 358–382.

Rabe, Barry G., and Rachel L. Hampton. 2015. Taxing Fracking: The Politics of State Severance Taxes in the Shale Era. *Review of Policy Research* 32 (4): 389–412.

Ramseur, Jonathan L., and James E. McCarthy. 2015. *EPA's Clean Power Plan: Highlights of the Final Rule*. Washington, DC: Congressional Research Service.

Raymond, Leigh. 2003. *Private Rights in Public Resources: Equity and Property Allocation in Market-Based Environmental Policy*. Washington, DC: Resources for the Future Press.

Raymond, Leigh, and Timothy N. Cason. 2011. Can Affirmative Motivations Improve Compliance in Emissions Trading Programs? *Policy Studies Journal* 39 (4): 659–678.

Raymond, Leigh, and Laura U. Schneider. 2014. Personal Moral Norms and Attitudes toward Endangered Species Policies on Private Land. *Conservation and Society* 12 (1): 1–15.

Raymond, Leigh, S. Laurel Weldon, Daniel Kelly, Ximena B. Arriaga, and Ann Marie Clark. 2014. Making Change: Norm-Based Strategies for Institutional Change to Address Intractable Problems. *Political Research Quarterly* 67 (1): 197–211.

Redefining Progress. 2004. Public Comment Allowance Allocation, September. http://www.rggi.org/design/history/public_comments (accessed November 25, 2015).

Reed, Stanley, and Mark Scott. 2013. "In Europe, Paid Permits for Pollution Are Fizzling." *New York Times*, April 21. http://www.nytimes.com/2013/04/22/business/ energy-environment/europes-carbon-market-is-sputtering-as-prices-dive.html (accessed December 1, 2015).

Regional Greenhouse Gas Initiative (RGGI). 2003. RGGI Subgroup on Model Rule Development: Model Rule Outline and Identification of Key Policy Decisions, Draft, December 12. http://www.rggi.org/docs/modelruleoutline.pdf (accessed November 23, 2015).

Regional Greenhouse Gas Initiative (RGGI). 2004. Regional Greenhouse Gas Initiative ("RGGI") Stakeholder Group Outline of Key Policy Issues for April 2 Meeting. On file with author.

Regional Greenhouse Gas Initiative (RGGI). 2005. Memorandum of Understanding. http://www.rggi.org/docs/mou_final_12_20_05.pdf (accessed November 28, 2015).

Regional Greenhouse Gas Initiative (RGGI). 2006. Public Review Model Rule Draft, March 23. http://www.rggi.org/docs/public_review_draft_mr.pdf (accessed November 28, 2015).

Regional Greenhouse Gas Initiative (RGGI). 2007. Model Rule, Final with Corrections, January 5. http://www.rggi.org/docs/model_rule_corrected_1_5_07.pdf (accessed November 28, 2015).

Regional Greenhouse Gas Initiative (RGGI). 2012. Regional Investment of RGGI CO_2 Allowance Proceeds, November 2011. On file with author.

Regional Greenhouse Gas Initiative (RGGI). 2013a. RGGI States Recommend That EPA Support Flexible Market-Based Carbon Pollution Programs. Press release, December 2. http://www.rggi.org/docs/PressReleases/PR120213_EPAComments_Final.pdf (accessed December 1, 2015).

Regional Greenhouse Gas Initiative (RGGI). 2013b. Twenty-First Auction Marks Five Years of Success for RGGI. Press release, September 6. http://www.rggi.org/docs/Auctions/21/PR090613_Auction21.pdf (accessed November 28, 2015).

Regional Greenhouse Gas Initiative (RGGI). 2013c. 2012 Program Review: Summary of Recommendations to Accompany Model Rule Amendment, February. http://www.rggi.org/docs/ProgramReview/_FinalProgramReviewMaterials/Recommendations_Summary.pdf (accessed December 1, 2015).

Regional Greenhouse Gas Initiative (RGGI). 2014a. CO_2 Allowances Sold at $4.00 at 23rd RGGI Auction. Press release, March 7. http://www.rggi.org/docs/Auctions/23/PR030714_Auction23.pdf (accessed December 1, 2015).

Regional Greenhouse Gas Initiative (RGGI). 2014b. RGGI States Make Major Cuts to Greenhouse Gas Emissions from Power Plants. Press release, January 13. http://www.rggi.org/docs/PressReleases/PR011314_AuctionNotice23.pdf (accessed December 1, 2015).

Regional Greenhouse Gas Initiative (RGGI). 2015a. About the Regional Greenhouse Gas Initiative (RGGI). http://www.rggi.org/docs/Documents/RGGI_Fact_Sheet.pdf (accessed November 22, 2015).

Regional Greenhouse Gas Initiative (RGGI). 2015b. Annual Report on the Market for RGGI CO_2 Allowances: 2014. http://rggi.org/docs/Market/MM_2014_Annual_Report.pdf (accessed December 6, 2015).

Regional Greenhouse Gas Initiative (RGGI). 2015c. RGGI Auction Results. http://www.rggi.org/market/co2_auctions/results (accessed November 10, 2015).

Regional Greenhouse Gas Initiative (RGGI). 2015d. Investments of RGGI Proceeds through 2013, April. http://www.rggi.org/docs/ProceedsReport/Investment-RGGI-Proceeds-Through-2013.pdf (accessed November 22, 2015).

Regional Greenhouse Gas Initiative Staff Working Group (RGGI SWG). 2004a. Detailed Agenda with Notes, SWG Meeting, August 18–19. On file with author.

Regional Greenhouse Gas Initiative Staff Working Group (RGGI SWG). 2004b. Meeting Agenda, January 15–16. On file with author.

Regional Greenhouse Gas Initiative Staff Working Group (RGGI SWG). 2004c. Meeting Agenda, February 26–27. On file with author.

Regional Greenhouse Gas Initiative Staff Working Group (RGGI SWG). 2005a. Agenda, Agency Heads Conference Call, February 18. On file with author.

Regional Greenhouse Gas Initiative Staff Working Group (RGGI SWG). 2005b. Memorandum to RGGI Agency Heads, Re: Public Benefits Set-Aside and Complementary Energy Policy Recommendations, June 24. On file with author.

Regional Greenhouse Gas Initiative Staff Working Group (RGGI SWG). 2005c. Memorandum to RGGI Agency Heads, Re: Revised Staff Working Group Proposal, August 24. On file with author.

Regional Greenhouse Gas Initiative Staff Working Group (RGGI SWG). 2005d. Memorandum to RGGI Agency Heads, Re: Staff Working Group "Straw" Proposal, June 22. On file with author.

Regional Greenhouse Gas Initiative (RGGI) Stakeholder Meeting Summary. 2004a. April 2. On file with author.

Regional Greenhouse Gas Initiative (RGGI) Stakeholder Meeting Summary. 2004b. June 24. On file with author.

Regional Greenhouse Gas Initiative (RGGI) Stakeholder Meeting Summary. 2005a. April 6. On file with author.

Regional Greenhouse Gas Initiative (RGGI) Stakeholder Meeting Summary. 2005b. May 19. On file with author.

Regional Greenhouse Gas Initiative (RGGI) Stakeholder Meeting Summary. 2005c. September 21. On file with author.

Regional Greenhouse Gas Initiative (RGGI) Stakeholder Meeting Summary. 2006. May 2. On file with author.

Regional Greenhouse Gas Initiative (RGGI) Stakeholder Workshop on Allowance Apportionment and Allocation. 2004a. Meeting Summary, October 14. http://www.rggi.org/docs/allocation_summary_10_28_04.pdf (accessed November 28, 2015).

Regional Greenhouse Gas Initiative (RGGI) Stakeholder Workshop on Electricity Markets, Reliability, and Planning. 2004b. Topic Session 1—Issues Relating to Conventional Power Supplies: Meeting Agenda, November 29. http://www.rggi.org/docs/topic_session_1_11_29_04.pdf (accessed November 28, 2015).

Reich, Charles A. 1964. The New Property. *Yale Law Journal* 73:733–787.

Ress, Dave. 2014. "A Cap-and-Trade Proposal for Virginia." *Virginia Daily Press*, December 22. http://www.dailypress.com/news/politics/shad-plank-blog/dp-virginia-politics-a-capandtrade-proposal-for-virginia-20141222-post.html (accessed December 1, 2015).

Richerson, Peter J., and Joseph Henrich. 2009. Tribal Social Instincts and the Cultural Evolution of Institutions to Solve Collective Action Problems. Paper presented in the Workshop on Context and the Evolution of Mechanisms for Solving Collective Action Problems, Indiana University, Bloomington.

Rio, Robert. 2011. Interview, November 17.

Risse, Thomas. 2000. "Let's Argue!": Communicative Action in World Politics. *International Organization* 54:1–39.

Rokeach, Milton. 1973. *The Nature of Human Values*. New York: Free Press.

Rudd, Kevin. 2013. Prime Minister's Press Conference. Townsville, AU, July 16. http://parlinfo.aph.gov.au/parlInfo/search/display/display.w3p;query=Id%3A%22media%2Fpressrel%2F2597362%22 (accessed December 5, 2015).

Ruddock, Robert. 2012. Interview, February 8.

Samuelsohn, Darren. 2010. "Senate Bill Allocation Fight Expected to Go Down to the Wire." *E&E Daily*, March 26. http://www.eenews.net/stories/89186 (accessed December 4, 2015).

Sandler, Mike. 2012. "The Birth of Carbon Pricing and Delivering California's First 'Climate Dividend.'" *Huffington Post Green*, December 5. http://www.huffingtonpost.com/mike-sandler/california-carbon-pricing_b_2205089.html (accessed December 4, 2015).

Santelli, Dan. 2006. RFF RGGI Auction Workshop Presentation. Presentation to the RGGI Allowance Auction Workshop, July 20. http://www.rggi.org/design/history/topical_workshops (accessed November 28, 2015).

Savitt, James. 2006. The RGGI Auction Allocation Process: Concerns and Recommendations. Presentation to the RGGI Allowance Auction Workshop, July 20.

http://www.rggi.org/design/history/topical_workshops (accessed November 28, 2015).

Sax, Joseph L. 1970. The Public Trust Doctrine in Natural Resource Law: Effective Judicial Intervention. *Michigan Law Review* 68:471–566.

Schmalensee, Richard, and Robert N. Stavins. 2012. The SO$_2$ Allowance Trading System: The Ironic History of a Grand Policy Experiment. Resources for the Future Discussion Paper 12–44. http://www.rff.org/files/sharepoint/WorkImages/Download /RFF-DP-12-44.pdf (accessed November 18, 2015).

Schmidt, Vivien A. 2008. Discursive Institutionalism: The Explanatory Power of Ideas and Discourse. *Annual Review of Political Science* 11:303–326.

Schneider, Anne L., and Helen Ingram. 1993. Social Construction of Target Populations: Implications for Politics and Policy. *American Political Science Review* 87 (2): 334–347.

Schneider, Anne L., and Helen Ingram. 1997. *Policy Design for Democracy*. Lawrence: University of Kansas Press.

Schneider, Anne L., Helen Ingram, and Peter DeLeon. 2014. Democratic Policy Design: Social Construction of Target Populations. In *Theories of the Policy Process*, ed. Paul A. Sabatier and Christopher M. Weible, 105–149. Boulder, CO: Westview Press.

Schneider, Anne L., and Mara Sidney. 2009. What Is Next for Policy Design and Social Construction Theory? *Policy Studies Journal* 37 (1): 103–119.

Seidman, Nancy. 2014. Interview, January 2.

Sekar, Samantha, Clayton Munnings, and Dallas Burtraw. 2014. Capitalising on Carbon Revenues. In *GHG Market Report 2014: Markets Matter*. http://www.ieta .org/resources/Resources/GHG_Report/2014/ieta%202014%20ghg%20report.pdf (accessed December 1, 2015).

"Seven Northeast States Move Ahead with Plan to Limit Greenhouse Gases." 2005. *Global Power Report*, December 22.

Shaw, Karena. 2011. Climate Deadlocks: The Environmental Politics of Energy Systems. *Environmental Politics* 20 (5): 743–763.

Sheehan, Denise. 2012. Interview, September 13.

Shobe, William. 2002. Auction Analysis Memo, May 16. On file with author.

Shobe, William. 2006. Allowances for Sale: Virginia's NO$_x$ Allowance Auction. Presentation at the RGGI Allowance Auction Workshop, July 20. http://www.rggi.org/ design/history/topical_workshops (accessed November 20, 2015).

Shobe, William. 2013a. Interview, July 30.

Shobe, William. 2013b. Personal communication, August 8.

Sikkink, Kathryn. 2011. *The Justice Cascade: How Human Rights Prosecutions Are Changing the World.* New York: W. W. Norton and Company.

Skjaerseth, Jon B., and Jorgen Wettestad. 2008. *EU Emissions Trading: Initiation, Decision-Making, and Implementation.* Burlington, VT: Ashgate.

Skjaerseth, Jon B., and Jorgen Wettestad. 2010. Fixing the EU Emissions Trading System? Understanding the Post-2012 Changes. *Global Environmental Politics* 10 (4): 101–123.

Skocpol, Theda. 2013. Naming the Problem: What It Will Take to Counter Extremism and Engage Americans in the Fight against Global Warming. Paper presented at the Politics of America's Fight against Global Warming, Columbia University, New York, February 14.

Sliwinski, Rob. 2013. Interview, October 17.

Sniderman, Paul M., and Sean M. Theriault. 2004. The Structure of Political Argument and the Logic of Issue Framing. In *Studies in Public Opinion: Attitudes, Non-Attitudes, Measurement Error, and Change,* ed. William E. Saris and Paul M. Sniderman, 133–165. Princeton, NJ: Princeton University Press.

Snyder, Jared. 2006. Preliminary Oral Comments of New York State Attorney General Eliot Spitzer on the Allocation of Carbon Dioxide Allowances Pursuant to the Regional Greenhouse Gas Initiative Cap-and-Trade Program. Presentation at the RGGI Stakeholder Meeting, May 2. On file with author.

Spolar, Matthew. 2011. "Lynch Vetoes RGGI Repeal." *Concord Monitor,* July 7.

Sripada, Chandra, and Stephen Stich. 2007. A Framework for the Psychology of Norms. In *The Innate Mind: Culture and Cognition,* ed. Peter Carruthers, Stephen Laurence, and Stephen Stich, 280–301. New York: Oxford University Press.

Stavins, Robert N. 1998. What Can We Learn from the Grand Policy Experiment? Lessons from SO_2 Allowance Trading. *Journal of Economic Perspectives* 12 (3): 69–88.

Stavins, Robert. 2012. Low Prices a Problem? Making Sense of Misleading Talk about Cap-and-Trade in Europe and the USA. An Economic View of the Environment (blog), April 25. http://www.robertstavinsblog.org/2012/04/25/low-prices-a-problem -making-sense-of-misleading-talk-about-cap-and-trade-in-europe-and-the-usa/ (accessed December 1, 2015).

Stigler, George J. 1972. The Theory of Economic Regulation. *Bell Journal of Economics and Management Science* 2 (1): 3–21.

Stone, Deborah. 2002. *Policy Paradox: The Art of Political Decision Making.* Rev. ed. New York: W. W. Norton and Company.

Stratton, Jessie. 2014. Interview, June 24.

Streeck, Wolfgang, and Kathleen Thelen, eds. 2005. *Beyond Continuity: Institutional Change in Advanced Political Economies*. Oxford: Oxford University Press.

Sunstein, Cass R. 2002. *Risk and Reason*. Cambridge: Cambridge University Press.

Svendsen, Gert Tinggaard. 1999. U.S. Interest Groups Prefer Emissions Trading: A New Perspective. *Public Choice* 101 (1–2): 109–128.

Svenson, Eric. 2011. Interview, December 6.

Swedish Environmental Protection Agency. 2010. *Baltic Survey: A Study in the Baltic Sea Countries of Public Attitudes and Use of the Sea*. Stockholm.

Tanzler, Dennis, and Sibyl Steuwer. 2009. *Cap and Invest: Why Auctioning Gains Prominence in the EU's Emissions Trading Scheme*. Washington, DC: Heinrich Boll Foundation. http://us.boell.org/2009/06/01/cap-and-invest (accessed December 1, 2015).

Taylor, Lenore. 2014. "Australia Kills Off Carbon Tax." *Guardian*, July 16. http://www.theguardian.com/environment/2014/jul/17/australia-kills-off-carbon-tax (accessed December 5, 2015).

Thelen, Kathleen. 2004. *How Institutions Evolve*. Cambridge: Cambridge University Press.

Tierney, Sue. 2005. RGGI Markets Workshop, 11-30-04: Summary and Key "Take-Aways." Presentation to the RGGI Stakeholder Meeting, February 16. On file with author.

Tietenberg, Thomas H. 1985. *Emissions Trading: An Exercise in Reforming Pollution Policy*. Washington, DC: Resources for the Future Press.

Tietenberg, Thomas H. 2010. Cap-and-Trade: The Evolution of an Economic Idea. *Agricultural and Resource Economics Review* 39 (3): 359–367.

Traxler, Christian, and Joachim Winter. 2012. Survey Evidence on Conditional Norm Enforcement. *European Journal of Political Economy* 28 (3): 390–398.

Tripp, James T. B., and Daniel J. Dudek. 1989. Institutional Guidelines for Designing Successful Transferable Rights Programs. *Yale Journal on Regulation* 6:369–391.

Truman, David B. 1951. *The Governmental Process*. New York: Alfred A. Knopf.

Tversky, Amos, and Daniel Kahneman. 1981. The Framing of Decisions and the Psychology of Choice. *Science* 211 (4481): 453–458.

Tyler, Tom R. 2006. *Why People Obey the Law*. Princeton, NJ: Princeton University Press.

Upton, John. 2015. Obama Just Created a Carbon Cap-and-Trade Program. *Climate Central*, August 4. http://www.climatecentral.org/news/obama-just-created-a-carbon -cap-and-trade-program-19309 (accessed December 5, 2015).

US Code of Federal Regulations. 2013. NO$_x$ Allowance Allocations. 40 Code of Federal Regulations, sec. 96.42. http://www.gpo.gov/fdsys/pkg/CFR-2013-title40-vol22/ pdf/CFR-2013-title40-vol22-sec96-42.pdf (accessed November 20, 2015).

US Environmental Protection Agency (EPA). 2005. *Draft Report: State Set-Aside Programs for Energy Efficiency and Renewable Energy Projects under the NO$_x$ Budget Trading Program: A Review of Programs in Indiana, Maryland, Massachusetts, Missouri, New Jersey, New York, and Ohio.* Washington, DC.

US Environmental Protection Agency (EPA). 2009. *Acid Rain and Related Programs: 2007 Progress Report.* Washington, DC.

U.S. Environmental Protection Agency (EPA). 2015a. Final Rule, Carbon Pollution Emissions Guidelines for Existing Stationary Sources: Electric Utility Generating Units, EPA-HQ-OAR-2013–0602, August 3. .

US Environmental Protection Agency (EPA). 2015a. Fact Sheet: Clean Power Plan Overview. http://www.epa.gov/cleanpowerplan/fact-sheet-clean-power-plan-over view (accessed January 11, 2016).

US Environmental Protection Agency (EPA). 2015b. Fact Sheet: The Clean Power Plan by the Numbers, August 3. http://www.epa.gov/cleanpowerplan/fact-sheet -clean-power-plan-numbers (accessed January 11, 2016).

US Environmental Protection Agency (EPA). 2015c. Federal Plan Requirements for Greenhouse Gas Emissions from Electric Utility Generating Units Constructed on or before January 8, 2014; Model Trading Rules. Docket EPA-HQ-OAR-2015–0199. https://www.gpo.gov/fdsys/pkg/FR-2015-10-23/pdf/2015-22848.pdf (accessed January 11, 2016).

US Environmental Protection Agency (EPA). 2015d. Final Rule, Carbon Pollution Emission Guidelines for Existing Stationary Sources: Electric Utility Generating Units. EPA-HQ-OAR-2013–0602, August 3. https://www.gpo.gov/fdsys/pkg/FR-2015 -10-23/pdf/2015-22842.pdf (accessed January 11, 2016).

US Senate Committee on Environment and Public Works. 2001. Hearings: Clean Air Act Oversight Issues, April 5.

Virginia Department of Planning and Budget (VA DPB). 2001. Economic Impact Analysis, 9 VAC 5–140–10 et seq.—Regulations for Emissions Trading, Department of Environmental Quality, March 14. http://townhall.virginia.gov/L/GetFile.cfm?Fil e=C:%5CTownHall%5Cdocroot%5C1%5C761%5C1089%5Cdeq1089.pdf (accessed November 20, 2015).

Virginia Department of Planning and Budget (VA DPB). 2002. Auctioning Allowances under the NO_x SIP Call: Briefing Outline for Secretary Bennett, March 21. On file with author.

Weible, Christopher M., Paul A. Sabatier, and Kelly McQueen. 2009. Themes and Variations: Taking Stock of the Advocacy Coalition Framework. *Policy Studies Journal* 37 (1): 121–140.

Weldon, S. Laurel. 2006. Inclusion, Solidarity, and Social Movements: The Global Movement on Gender Violence. *Perspectives on Politics* 4 (1): 55–74.

Western Climate Initiative (WCI). 2008a. Design Recommendations for the WCI Regional Cap-and-Trade Program, September 23. http://www.westernclimateinitiative.org/the-wci-cap-and-trade-program/design-recommendations (accessed December 4, 2015).

Western Climate Initiative (WCI). 2008b. July 23 Draft Design Recommendations Comments. http://www.westernclimateinitiative.org/draft-design-recommendation-comments (accessed December 4, 2015).

Weyland, Kurt. 2005. Theories of Policy Diffusion: Lessons from Latin American Pension Reform. *World Politics* 57 (2): 262–295.

White, Jeremy B. 2014. "California Republicans Seek Cap-and-Trade Exemption for Fuels." *Sacramento Bee*, December 1. http://www.sacbee.com/news/politics-government/capitol-alert/article4226376.html (accessed December 4, 2015).

Wiessner, Polly. 2009. Experimental Games and Games of Life among the Ju/'Hoan Bushmen. *Current Anthropology* 50 (1): 133–138.

Wilson, James Q. 1989. Interests. In *Bureaucracy*, by James Q. Wilson, 72–89. New York: Basic Books.

Wood, Mary Christina. 2012. Atmospheric Trust Litigation across the World. In *Fiduciary Duties and the Atmospheric Trust*, ed. Ken Coghill, Tim Smith, and Charles Sampford, 99–163. Burlington, VT: Ashgate.

Wood, Mary Christina. 2014. *Nature's Trust: Environmental Law for a New Ecological Age*. New York: Cambridge University Press.

Young, H. Peyton. 1994. *Equity: In Theory and Practice*. Princeton, NJ: Princeton University Press.

Younger, Mark. 2004. CO_2 Allowance Allocation in Regional Greenhouse Gas Initiative. Presentation to the RGGI Allocations Workshop, October 14. http://www.rggi.org/docs/younger_pres_10_14_04.pdf (accessed November 25, 2015).

Younger, Mark. 2005. Identifying the Appropriate Allocation Approach. Presentation to the RGGI Stakeholder Meeting, April 6. On file with author.

Zafonte, Matthew, and Paul A. Sabatier. 2004. Short-Term versus Long-Term Coalitions in the Policy Process: Automotive Pollution Control, 1963–1989. *Policy Studies Journal* 32 (1): 75–107.

Zerbe, Richard O., and C. Leigh Anderson. 2001. Culture and Fairness in the Development of Institutions in the California Gold Fields. *Journal of Economic History* 61 (1): 114–143.

Index

American and Comparative Environmental Policy

Sheldon Kamieniecki and Michael E. Kraft, series editors